SAFARI
JOURNEY TO THE END

Wolfgang Bell,
Safari Consultant,
R.R. 1, Bancroft, Ont. K0L 1C0
Tel/Fax 1-613-332-1001

SAFARI
JOURNEY TO THE END

DAVE TAYLOR

THE BOSTON MILLS PRESS

*To Liza and Ashley, who stayed home, and to their mom,
Anne, who stayed with them.*

Canadian Cataloguing in Publication Data

Taylor, J. David, 1948-
  Safari: journey to the end

Includes bibliographical references.
ISBN 1-55046-016-1

1. Natural history - Kenya. 2. Zoology - Kenya.
3. National parks and reserves - Kenya. 4. Safaris - Kenya.
5. Kenya - Description and travel - 1981-
I. Title.

QH195.K4T38 1990   508.6762   C90-093307-0

Published by:
THE BOSTON MILLS PRESS
132 Main Street
Erin, Ontario
N0B 1T0
(519) 833-2407
Fax: (519) 833-2195

American Association
for State and Local History
Award of Merit

Winners of the
Heritage Canada
Communications Award

Typography and design by Lexigraf, Tottenham
Cover design by Gill Stead, Guelph
Printed by Khai Wah Litho PTE Limited, Singapore

The publisher wishes to acknowledge the financial assistance of The Canada Council,
the Ontario Arts Council and the Office of the Secretary of State.

Author's Note:
The photographs in this book were taken with Canon cameras and lenses. I am grateful
for this company's support. All of the slides were taken with Kodachrome 64 film.

Contents

Foreword

Why the End is Kenya!

Kenya is one of the world's most beautiful countries, blessed with a remarkable diversity in scenery: snow-capped mountains to plains, forests, grassland to semi-desert. The great scenery is complemented by an abundant variety of wildlife.

Kenya's major parks — Tsavo, Amboseli, Masai-Mara, Meru and Samburu — all feature not less than fifty species of wildlife. The Amboseli National Park is considered by many as the place to see lions, elephants, buffalo and rhino. The Masai-Mara, on the other hand, offers one of nature's most spectacular sights, with millions of wildebeest and zebra migrating across the plains. The Mara also offers the opportunity for one to see some of the rare and less frequently seen animals.

David Taylor's book, *Journey To The End*, which is organized and presented in diary form, contains an account of real experiences and warm memories from his visit to Kenya. The author presents vivid descriptions of his observations at Amboseli and Masai-Mara, and supports these with magnificent photographs which highlight the attractions of animal life at the two parks.

This book should appeal to animal lovers, ardent birdwatchers and travel agents who value sending their clients to memorable destinations.

The trip to Kenya is a visit of a lifetime. It is a journey to The End. It is where the grass and the sky meet.

Peter Maragia Nyamweya
Kenya's High Commissioner to Canada

Simon Mutunga and the author.   - Jim Markou

DAVE TAYLOR is a teacher for the Peel Board of Education and a freelance writer and photographer specializing in natural history subjects.

He is a field editor for *Ontario Out of Doors* Magazine, where his work appears regularly. He also contributes to *Landmarks* magazine and a number of other publications across North America.

He is the author of *Ontario's Wildlife* which won the 1988 Nissan Book Award for best outdoor book in Canada.

Acknowledgements

The key to the success of the trip was our driver, Simon Mutunga, whose insight and knowledge of wildlife and the parks' terrain proved invaluable. Without him we would have missed many photo opportunities, for often ours was the only vehicle present when things happened.

Having along a brother-in-law on such a trip may seem an odd choice, but when he is a talented photographer as well, it makes a good deal of sense. Jim Markou and I have worked together on several such trips, and while the competition to market our material afterward may be stiff, on these trips we feed off of each other's knowledge, talents and intuitions, improving both our results. To be good friends as well doesn't hurt, and I thank him.

We, and for that matter the world, owe the government and people of Kenya a debt of gratitude for their wisdom in setting aside such places as Meru and the Mara. We hope they will continue to see the wisdom in their approach and strengthen their resolve to improve protection measures even further.

African Safari Club, through their Canadian representative, Wolfgang Bell, kindly organized and provided additional support for us while in Africa. In addition to thanking Wolfgang and Simon, I should particularly like to express my appreciation to Frank and Marion Neugebauer for their kindness and help during our stay at the Mara Buffalo Camp.

Alex Tilley of Tilley Endurables supplied our safari clothes, which proved to be better than promised.

All of the photos were taken with Canon t-90 cameras. The lenses used were: 200 2.8, 80-200 Zoom 4L, 300 4.5L and a 500 4.5L. The film used was Kodachrome 64.

I must caution the reader, however, that while I could afford only two weeks, the safari Jim and I planned was not the usual packaged trip. We arranged through the African Safari Club to have our own car and driver, and this allowed us the freedom to spend several hours watching wildebeest, wild dogs or lions. Timewise, this luxury is something the typical tour group cannot afford. These groups tend to move from place to place, species to species, more rapidly than I would wish to go. On the other hand, spending an hour or two watching sleeping lions on the off chance they might do something can be boring indeed, especially if the sun sets before they get up!

The option of mode of travel, group or private, is available to anyone, of course, but I would suggest our approach only if the traveller is very serious about photography and is prepared to invest the effort at the expense of the more relaxed and enjoyable pace of group travel. Group trips are considerably less expensive and provide excellent opportunities to see a variety of Kenya's wildlife as well.

As a footnote, I should add that all of the photos in this book were taken within two weeks, August 6 to 20, and all are in the wild. You will no doubt notice there are no shots of leopards. We saw this one species only once and never did succeed in photographing it — one of the reasons I'll go back.

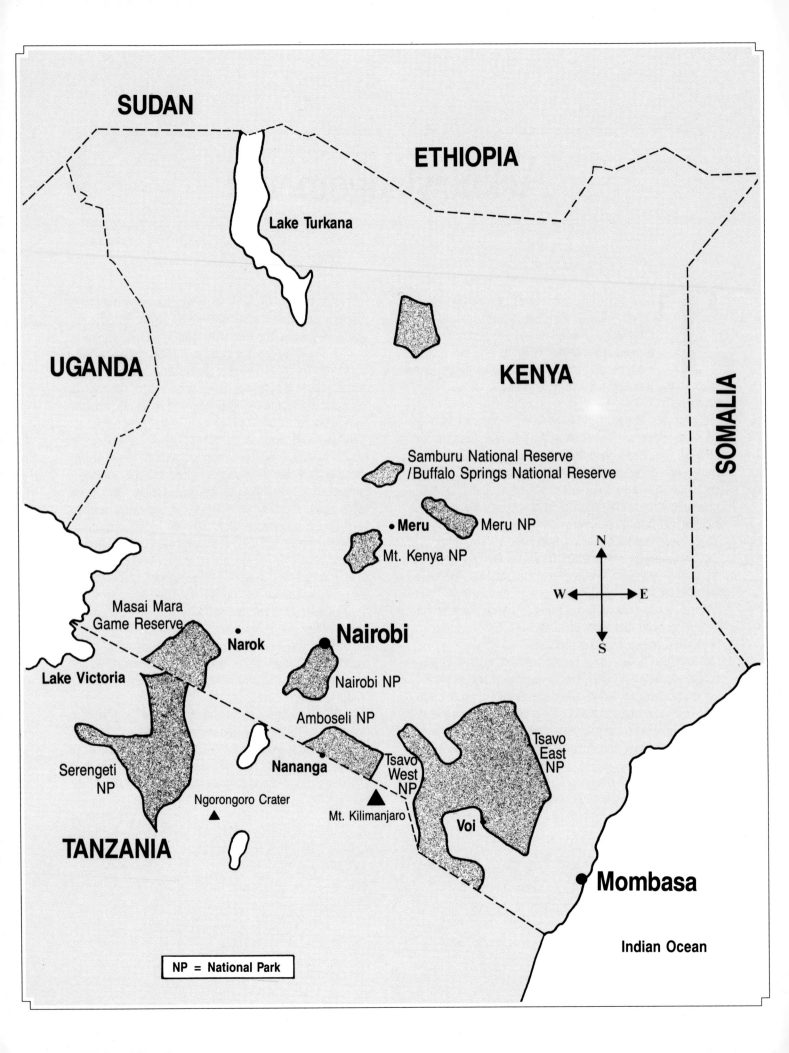

Introduction

"The Maasai have a saying that where the grass and the sky meet, the name of that place is The End."

Those words are from Alan Root's film of the Serengeti's wildebeest migration, *The Year of the Wildebeest*, the film that rekindled my boyhood interest in the Dark Continent. A decade passed between the time I first saw that remarkable film and the moment I realized that its images and words would not fade. I wanted to see for myself that place the Maasai call The End. I wanted to see the massive herds of wildebeest on the move and the drama that unfolds around them.

I achieved that goal one recent August in Kenya. Technically, the Serengeti plain ends at the border between Tanzania and Kenya, but that border is man-made and wildlife don't respect such artificial boundaries. For them the plains don't end until they reach the hills and escarpments in Kenya's Masai-Mara National Reserve. It was to this location my brother-in-law and fellow photographer, Jim Markou, and I journeyed.

We spent two solid weeks on safari, and this book is a result of those experiences. I took some liberties with time in the narration, condensing several days' events into one, especially in the chapter on the Mara, but the events happened as I describe them. They reflect what I saw and photographed, or in some cases learned from other travellers who were on the scene before we arrived.

The background information on the various animals is based on thorough research and is as accurate as possible. A bibliography of sources is provided for the serious reader.

I have not dealt with the important problems facing Africa's wildlife in any great depth. My hope is that this book, through its photos and text, will serve to convince the reader that such animals are important to preserve.

Ultimately, my hope is that some who read this book will go on safari. These areas are allowed to exist because they bring in tourist dollars. The more people who go, the better the chances of these precious places continuing to survive.

I shall return. I wasn't certain I would until that second day in the Mara, when we met up with another bunch of tourists. We had driven for a couple of hours through the grassland and had seen herds of a hundred or so wildebeest, but nothing like the numbers shown in Root's film, and then we met this other safari vehicle.

The fellow in the other car was beaming with pleasure. I asked him what he had seen. "Thousands of animals. Everywhere, animals for as far as you can see," he said. "I've been to the end of the world."

A few minutes later I went there too.

The Dry North - Days 1 to 3

Meru National Park
1,813 sq. km. (700 sq. mi.)

Samburu National Reserve
104 sq. km. (40 sq. mi.)

Buffalo Springs National Reserve
194 sq. km. (75 sq. mi.)

The land is dry here, robbed of its moisture by the soaring bulk of Mount Kenya scarcely 100 kilometres (60 miles) away. Yet it is that same mountain, along with those in the Aberdares Range, that is the source for most of the rivers rushing through the area, and one in particular, the Uaso Nyiro River, which for 32 of its kilometres forms the boundary between Samburu and Buffalo Springs reserves.

A little further south and east is Meru National Park, but it too is a product of the same forces that influence the other parks in this chapter, and for purposes of this narration it is simpler to deal with all three as if they represented one ecosystem.

The landscape in all three consists of gently rolling hills interspersed with plains. The vegetation is in places sparse, almost nonexistent at times, until underground springs, rivers or rainfall allow first grasses and then trees to take root. Even these wither and yellow in the driest of times, and then the only green to be seen marks the course of the river, where trees form a dense margin providing wildlife both shelter and places for ambush.

Of the three areas, Meru is the wettest and boasts three major rivers: the Rojerwero, the Ura and the Tana. It is also one of the better-known reserves in the world, thanks to a former resident whose life and death achieved world fame. It was here that much of the story of the lioness Elsa (*Born Free*) took place, and her memory is everywhere

Reticulated Giraffe, Buffalo Springs

in this park, including such attractions as "Elsa's Camp."

Sadly, her legacy is not readily visible in the form of living lions. There are lions here and in Samburu - Buffalo Springs, but sightings are not as common as they once were. Poaching appears to be the main reason the surviving cats have adopted a more cautious approach to life in these reserves than have the lions further to the south.

Cheetahs are also found here, and being strictly diurnal hunters, they are easier to see, but they too are quick to disappear into the scrub brush of the region.

Easily seen and often encountered is the magnificent reticulated giraffe. Once considered a separate species of giraffe because of its bold markings, it has been reclassified as a subspecies based on its ability to interbreed with other types of giraffes. Like all giraffes, they feed on the leaves of several trees and bushes, and may even limit the growth of these plants with their constant browsing.

Giraffes appear to have a very loose-knit social system. While they are often encountered in herds, these groups are in a constant state of change. Individual group members frequently leave to form new groups, with few if any members of the original group remaining. The advantages of this type of social system are still under research and no entirely satisfactory explanation has been put forth.

The shifting loyalties of the giraffe is in direct contrast to the territorial behaviour of Kirk's dik dik. Dik diks are among the smallest of the antelopes and are encountered quite often in this country. Only 36 to 41 centimetres (14 to 16 inches) tall, they are a creature of the underbrush, where they have developed a rather rabbit-like response to danger — dart, swerve and freeze! This method of avoiding predators is a common one among small bush-dwelling species and contrasts markedly with the run-for-your-life approach adopted by plains-dwelling prey.

Unlike either the rabbit or the giraffe, dik diks are decidedly monogamous. Once a male has selected a mate, the pair remain together for life. Both male and female will defend their territory against other dik diks, and it is possible for the same two dik diks to occupy the same small piece of brush for several years.

Along the boundary of that territory they will establish several middens, or dung heaps, that they will build up by constantly returning day after day, with both adults and offspring contributing. Local folktales offer the following explanation for these unusual piles.

One day, so the story goes, a dik dik was out walking when it tripped over the droppings of an elephant. This so infuriated the dik diks that they decided to pile their droppings up and trip the elephant, and so began the tradition.

Some scientists have suggested a possibility which is a little less fanciful. They propose that the middens are simply a way of marking borders between one family's home territory and another's. These middens can be seen throughout the dik dik's range, and this raises an interesting question. If the available space is all divided up, what does a young dik dik do for a territory?

In nature's scheme there is very little a young male dik dik can do. It will be chased off of its parents' territory and into some other's, where it will also be forced to leave. All this leaves the young and inexperienced animal tired, confused and vulnerable. In most cases it falls prey to one of the dik dik's many predators. (Females fare about the same.)

Its only hope is to stumble into a territory where one of the established adults has been killed. In most years this is a rare event, but occasionally flooding and/or diseases greatly reduce dik dik numbers, and then the young dik diks serve to replenish the population.

Another interesting antelope of the dry north is the long-necked gerenuk. Unlike the dik dik, which is found in all of Kenya's parks, the gerenuk is limited to only a few. This giraffe-antelope, as it is also called, is found in all three of the areas in this chapter but is nowhere more abundant than in Samburu and Buffalo Springs reserves. Like the dik dik, the male gerenuk establishes a home territory, but unlike the smaller antelope, gerenuks do not pair for life. The male will have several females throughout his range and will breed with them as they come into season. Young females may stay in this space, but juvenile males are not tolerated and are chased out.

These young bucks do not face the certain death that many dik dik young face. Because gerenuks have more available space than do the smaller antelope, it is not uncommon to see small bachelor groups of two or three males living on the edges of adjoining territories. These males also have a better chance of taking over additional territories because the rigours of breeding and chasing off rival males soon tires out the dominant male, making him vulnerable to attack by a rested challenger.

None of these species exists in great herds. The land here isn't rich enough to support them and bush living doesn't require as many eyes. As a result, when a predator charges, it is usually every animal for itself. A common behavioural pattern is that the more dense the tree cover, the smaller the group of animals.

It is not uncommon to encounter herds of buffalo, elephant, impala and Grant's gazelle, but the most spectacular antelope and largest to be seen in herds here is the Beisa oryx.

These horse-sized antelope are thought to have been the inspiration for the legend of the unicorn, and it is not uncommon to see an oryx with only one horn. Their long horns are among the most spectacular in the animal kingdom. Like all such horns, these are designed for battle with rival males. However, the female of the species has also developed the impressive horns, most likely as a weapon for warding off predators. Lions have been found impaled by these animals, and there are accounts of oryx successfully fighting off wild dog packs.

Newborn oryx are vulnerable to predators, often falling behind if the herd stampedes. Like many bush-dwelling antelope, the oryx's strategy for survival is simple: stay put and lie still. Because they almost entirely lack scent, such behaviour may save their lives unless the predator happens to trip over them. The strategy works because most animals, predator or prey, see movement far better than they see shapes.

Another rare species found only in Meru National Park has no such problem. The white rhino, being the largest of its kind, has little to fear from predators. White rhinos have been reintroduced to Meru from other parts of their range in Africa. The fossil record shows that the species once existed here, and in an effort to restore the white rhino to its natural habitat and save it from extinction, the government of Natal has been supplying breeding herds to several African countries. Meru's herd is one such example. Sadly, after this book was written a gang of poachers slaughtered the 5 white rhinos in Meru. They were the last of their species living on public land in the country. Hopefully they will be replaced in the near future.

Unlike its close cousin, the black rhino, the white rhino is a grazer. Its lips are "square," much like a cow's and are ideally suited for collecting enough grass to feed its massive body. The black rhino's lip is pointed and is used to grasp the twigs and leaves on which it browses. Their choice of food also affects their lifestyle. The white rhino is able to live in herds because there is generally more grass available than browse. The black rhino is most often encountered singly, in the company of its calf, or in a courting pair.

There is one large tree-eating animal found in these parks that does live in herds, and that is the elephant. The African elephant is the largest living land animal and as such has little to fear from predators. There is really little reason for them to live in groups. That they do is probably a measure of their intelligence and need to socialize.

The typical elephant herd consists of females and their calves and usually numbers less than 20. All of the adult females in the herd are friendly and know each other well. Female calves remain with the group even after maturity, and it is possible that all of the animals may be aunts, sisters and cousins.

Bull elephants leave the herd as they approach sexual maturity. The teenagers begin making advances on the adult females and are quickly repulsed, often violently. At first they hang about the herd on its edges, but gradually they wander off.

As adults, bulls and cows seldom occupy the same ranges. The exception being that adult males will from time to time visit a cow herd to see if there are any breeding females. Young females find these males exciting and often appear to flirt with them, but the older ones take their presence in stride and ignore them unless in heat.

Bulls over 35 come into a breeding behaviour called *musth* once a year, for three months. Although a bull is capable of breeding at the age of 12 and will court females whenever the opportunity occurs, musth is something more than just normal breeding behaviour. For the three months of musth, these dominant males — which may weigh more than twice that of the females — are extremely aggressive and undergo a sort of Jekyll and Hyde transformation. They also change physically, their temporal glands swelling and leaking a stream of liquid down the sides of their heads. Females find them stimulating and other males find them threatening, both factors which make successful breeding more likely.

The female herds, however, are the root of elephant society, and these units stay intact as males come and go. Occasionally two closely allied groups will meet up and spend some time together, and on even rarer occasions several such groups will gather to form a herd of several hundred individuals. Such gatherings are rare and are sights to be remembered and treasured.

Finally, one other oddity to be found in these parks is an animal that may or may not be a zebra. In all three reserves, two types of zebras may be seen. One, the common or Burchell's zebra, resembles in physical size and shape other zebras found throughout the African continent. There are some slight differences in the various subspecies' stripe patterns, but essentially they look the same.

The other zebra found here, the Grevy's zebra, has stripes too, but the stripes are much finer and don't extend onto the belly. It is not its stripes, however, that pose the problem of identification. This zebra simply doesn't look like a true zebra. It is bigger, its face is longer, and in general it looks more like a mule than a zebra; except for the stripes, that is. So is it a zebra or a zebra-like horse? It remains a question for scientists to decide.

Reserves like Samburu, Buffalo Springs, and Meru National Park provide some protection for these great beasts, and with growing concerns about conservation, it is to be hoped that they will continue to do so for some time to come.

A one tusk elephant.

Overleaf: Giraffes are found wherever there are trees, even in the dry semi-arid regions of Kenya.

A cow elephant pauses for a drink along the Uaso Nyiro River.

Two crocodiles warm themselves in the early morning light along the Uaso Nyiro River.
These animals can be found in most of Kenya's rivers and lakes.

Opposite: The Reticulated giraffe has bolder markings than its southern cousin the Masai giraffe.

Cheetah cubs rely on their mother not only for food but also for lessons on how to hunt.

A cheetah mother often leaves her cubs hidden in the brush while she goes out to hunt.

A Kirk's dik dik will live in the same area all its life if it is successful in finding a spot that is not already occupied by other dik diks.

Dik diks mate for life and chase even their own young from their territory often condemning them to death. Such apparently selfish behaviour appears to be more common in the animal world than first thought.

A male gerenuk is a strange looking antelope that seems to be part impala and part giraffe.

Male gerenuks seek out food on the edge of territories held by dominant males and await their turn to take over the harems.

Gerenuk are well adapted to feeding on trees. If their long necks don't reach they simply stand on their back legs.

The presence of Common waterbucks indicate that year round water must be available even in this dry land.

Beisa oryx escape the heat by seeking what little shade exists at midday.

A newborn Beisa oryx relies on lying still to escape detection by predators.

*Although oryx with their long horns have been known to kill lions with them,
the horns are primarily used as defence in rutting battles.*

Overleaf: White rhinos are herd animals largely because they are grazers.

A local waterhole is a popular destination for the elephants of Buffalo Springs National Reserve.

Their youngsters protected in the middle of the herd the elephants pass under some acacia trees dripping with weaver bird nests.

Opposite: Elephants can destroy trees with great ease as the broken limb attests.

Overleaf: A large herd of elephants in Meru, one of the world's great sights.

Whether a Grevy's zebra is a true zebra or merely a type of striped horse is still a matter of much debate.

A Grevy's zebra and her nearly full grown foal.

Common zebra are found in these parks too but they seldom associate with the Grevy's species.

The foothills of Mount Kenya make a striking background as a herd of zebra moves out onto the grassland to spend the night.

Striped ground squirrel

Agama lizard

White-breasted bustard

Crowned Cranes

Somali Ostrich

Redbilled Hornbill

Helmeted Guineafowl

Yellow-necked spurfowl

Marabou stork

Common zebras

Mountain Interlude - Days 4 and 5

Mount Kenya National Park
588 sq. km. (227 sq. mi.)

That the water hole isn't actually within the boundaries of Mount Kenya National Park is of little importance to the animals that use it daily and nightly. And it would appear that the large wooden building overlooking it is of even less importance, except of course to the troops of Sykes' monkeys that regularly search it for scraps of food and other interesting items.

The building is the Mountain Lodge, one of three such lodges to be found in Kenya. The other two are Treetops and The Ark. All three are built on "stilts" and all overlook an open area where water and salt may be found. Treetops and The Ark are in Aberdare National Park, while the Mountain Lodge is on the slopes of Mount Kenya.

Technically, the park which bears the mountain's name includes any part of the mountain above the 3,365 metre (11,034 foot) contour. The Mountain Lodge is sited a little above the 2,134 metre (7,000 foot) level in the dense Mount Kenya forest reserve. Here rain and mist nourish a splendid forest of soaring trees that hide a rich variety of wildlife.

The animals that come in to drink and lick the salt choose to ignore the Homo sapiens watching them. The observer is therefore able to watch the drama of the African forest unfold.

The most common large mammal here is the Cape buffalo. These aggressive animals are more than capable of fighting off any wild predator found in the forest. They plod through the mud searching for clumps of grass or moisture, and anything foolish enough to stand in their way is met with a cold, menacing stare.

Opposite: Elephant, Mount Kenya

Egyptian geese apparently do not bluff easily and are given to loud honking protests when too many buffalo descend on their pond. When the lumbering bullies get too close, the geese simply move off with even louder protests.

The nastier tempered buffalo are the males. Their horns may spread well over a metre, and unlike the water buffalo's horns, the Cape's horns meet in the centre of its forehead, forming a massive boss that neither lion nor bullet can penetrate. They have been known to attack and kill lions, and are generally considered to be the most dangerous animal in Africa. There are many accounts of lone bulls charging and severely damaging safari vehicles, often without apparent provocation!

Surly by nature, a bull buffalo backs down from nothing. When two such bulls meet, anything can happen, from outright battle to complete indifference, but, buffalo being buffalo, they usually stand and stare at each other, each sizing up the other. Sooner or later one of the bulls will either lose its nerve and glance away or the two will fight. Fights are uncommon because bulls establish an order of dominance. However there is often a lot of posturing before the weaker bull moves off.

Such stand-offs are a common occurrence around the mountain water hole. What is surprising is how often buffalo treat other creatures in the same manner. There are few animals that will stand their ground or challenge another species unless there is a very good reason to do so. Buffalo don't need the excuse of a newborn calf or a choice piece of grazing territory to challenge an intruder.

The black rhino, an animal which has a reputation for being aggressive, pretty much adopts a live and let live philosophy unless challenged or threatened, but when a buffalo meets up with a rhino, the buffalo just naturally takes offence. In this case the odds are about even. The rhino will outweigh the bull, but the buffalo has speed and a fairly well-armed head too. Such encounters, if pushed, usually end with the rhino chasing off the bull or with the buffalo finally coming to its senses and moving off. Most such encounters happen at night around the water hole, for Mount Kenya's rhinos are very cautious.

It would seem likely that the towering bulk of an elephant would give even the buffalo a reason for moving out of the way without challenge, but buffalo aren't about to give in to some wrinkled old pachyderm. Not without a challenge!

When the elephants arrive in the vicinity of the water hole, they do so in one of two ways. Loud trumpeting and much crashing of bush as they feed might announce their arrival or they might just suddenly appear with no sound or fanfare. To watch these grey creatures emerging out of the mist-shrouded forest is a magical experience. They come on quietly with frequent pauses by the lead cow to check out the situation. Young calves under six months don't stray far from their mothers. Even the larger ones follow when the herd moves, and they too come in to drink with a certain regality about them.

To get to the water they often must pass through a herd of buffalo that have bedded down in the meadow. Female buffalo and calves will move reluctantly out of the way, but not the bulls! Nothing, not even a whole herd of elephants, will budge them. At least that's what they'd like you to think.

The lead elephant cow will turn and flare her ears at the truculent buffalo bull. The rest of the herd may even move behind her and into the water hole, but she will watch the bull buffalo until he moves off or charges. Now there is no way even a big bull has much hope of beating an elephant and he knows it, so the outcome is a foregone conclusion. Still, appearances must be maintained, and it is only after an appropriate period of time that the bull backs down. The cow elephant then joins the rest of the herd and the little melodrama ends.

The show has its comedy too. Adolescent elephants are quick to size up the buffalo for what they are: bullies. They appear to take delight in terrorizing the heavy horned creatures and will chase them whenever they can. A stately cow elephant is one thing, a rampaging teenager is another, and the buffalo, their dignity gone, scatter before the onslaught of these cheeky grey devils. No one gets hurt and it is great fun to watch.

There are other watchers here too, and they are here for more serious business. The delicate bushbuck comes in to drink with considerably more care than do the buffalo, elephants and rhinos. A careless move and they could wind up as dinner for a hiding leopard.

Bush pigs also visit the water hole but only at night, as do their larger cousins, the giant forest hogs. These latter animals are nearly as large as the buffalo and are deadly foes when cornered. Their young, however, are the size of piglets and are easy prey, so the parents keep a careful watch over them.

The Sykes' monkeys seem almost careless when they arrive, but it is an illusion. Death hunts them from the air, the trees and the ground, and they know it. Their eyes are constantly shifting, heads jerking to attention, bodies ready to flee. To be less vigilant is to die.

Up in the trees another monkey is much more relaxed. The white-and-black colobus monkey is a treetop-dweller, and as such has fewer predators to worry about. These monkeys are capable of spectacular leaps — an escape plan that leaves even the agile leopard behind.

Leopards and other large predators are seldom seen at the water hole because they are so few in number. A much smaller predator is often sighted, however, and that is the black-tipped mongoose. It appears out of nowhere and moves rapidly around the pond searching out insects or small animals to eat.

Birds are abundant here and a variety of species come and go. Lappet-faced vultures come down to drink while hamerkoks prowl the grassy edge of the water hole. Mousebirds and a host of other smaller birds constantly flit back and forth while eagles soar over head, sometimes stooping on these other birds.

It is, all and all, an ideal spot to watch an African tableau entirely different from the ones to be seen on the plains or in scrub brush country.

Opposite:
Buffalo are cantankerous animals and none more so than when two bulls of equal size meet face to face. Eventually the facing bull backed down.

This huge bull is a formidable animal that even lions would avoid attacking.

Buffalo come to the salt lick not only for the minerals but for the rich grazing that exists around the waterhole.

A calf buffalo will stick close to its mother for protection and if it is a female it will likely stay in the same herd with her for its entire life.

Bull buffalo aren't afraid of anything it seems, including elephants!

When elephants arrive at the waterhole they move in as a group.

Young calves like this one stay close to their mothers.

Opposite: A charging teenage bull.

Elephants can drink several gallons of water at a time.

Elephants are one of the few animals to have four knees, something that comes in handy when trying to gouge out the mineral rich earth of the lick.

A young bull elephant terrorizes a bull buffalo.

A young male bushbuck approaches the waterhole cautiously.

Towards dark, giant forest hogs approach the waterhole.
These animals can be nearly as large as the Cape buffalo.

Opposite: Sykes' monkeys are excellent scroungers and have lost all fear of humans at the Lodge.

Black and white colobus monkey

Weaver birds

Mousebird

Black-tipped Mongoose

Hamerkok

Egyptian Goose

Introduction to the Grasslands - Days 5 and 6

Nairobi National Park
114 sq. km. (44 sq. mi.)

There is perhaps no finer way to be introduced to the principal characters of the African grassland than by spending a leisurely day or so exploring Nairobi National Park.

Created in 1946, the park still contains almost all the species associated with an African safari. The one obvious exception is the elephant, whose presence only 10 kilometres (6 miles) from the centre of Kenya's biggest city presents potential problems should the herd take up raiding local gardens. Therefore elephants are kept out.

Contrary to the impression left by travel brochures, the lion population is also controlled, so that the big cats are not as visible as one might expect. There are some lions in the park, and they go about their daily routine apparently unconcerned about their two-legged neighbours.

With herds of zebra, impala, eland and assorted other antelopes in the park, an ostrich would seem unlikely prey for a lion. The largest member of the bird family is fleet of foot and can easily outrun a lion, given half a chance. Even if cornered, its powerful feet have been known to shatter the big cat's jaw, so a lion is likely to give such a hunt a second thought. The ostrich's long neck and keen eyesight also make a successful stalk an improbable event, but sometimes dumb luck intervenes in the scheme of things.

On this particular morning a large male ostrich had little thought of predators. His bright red neck and legs identified him as a male bird belonging to the Masai species of ostrich. The Somali species has blue legs and neck and is found further north. Blue- or red-necked, the colour is at its most brilliant only during the breeding season, and that fact helps to explain how the lioness made

Opposite: Lioness, Nairobi National Park.

this unusual kill. The male ostrich was probably seeking females and was so wrapped up in the pursuit that he gave no thought to the clump of bushes growing in the shade of the old tree. The lioness that lay there wasn't hunting; she was resting, completely hidden in the underbrush. The male ostrich could not have seen her. He, however, was of immediate interest to the lioness.

When he passed by, the lioness charged. There was no time for the ostrich to kick or run. As evidenced by little visible sign of a struggle, death came quickly.

While the choice of victim was unusual, such hunts are daily events in this small park, almost all of which is within sight of the Nairobi skyline.

That the Kenyan government had the foresight in 1946 to preserve such an area is remarkable. That the growing populace continues to allow it to exist is a tribute to their commitment to both conservation and tourism. Agriculture and housing could both overwhelm this small park, but the need for tourist dollars seems to assure its continued existence.

Wildlife in Kenya is preserved as much for its economic value as for anything else. This may seem a callous reason for conservation, but with Kenya's growing population of over 25 million, any land set aside as parks or reserves must pay its own way. A million tourists a year justify such commitments, and for most of these visitors Nairobi National Park is their chance to see wild Africa.

It is also the best place in Kenya to see black rhino. The black rhino population is being decimated throughout its range in Africa, and within a few years it may cease to exist in the wild. Its undoing is its horn, which is valued as an aphrodisiac in the Far East and, more importantly, for the handles of daggers in North Yemen. Such horns can fetch well over a year's wages in Kenya shillings, and many poachers consider such profits worth the risk.

If Kenya is able to save her rhinos, it will be in the smaller parks like Nairobi that she will have her best chance. The big parks are too large and the park rangers too few to allow for truly effective anti-poaching units. It is easier to provide such protection in Nairobi National Park, and that fact alone explains why there are 25 rhinos here and perhaps only two or three in the entire Masai-Mara reserve, which is over ten times Nairobi's size!

Black rhinos are not the stupid, blundering beasts they are made out to be. Like elephants, they can be extremely dangerous in certain situations, but for the most part they go about their business in a quiet manner. Usually they are encountered alone, a behaviour pattern which sets them apart from the other rhinos found in Kenya, the white. Females with young and courting pairs are the exceptions to the rule.

It is true that rhinos have poor eyesight, and this may account for some of the "charges" they are noted for making. In order to see whatever it is that has caught their attention, they may run toward it for a better look. Most often, though, they raise their tails and make off for safer territory.

Natural enemies are confined to hyenas and lions, neither of which are much of a factor in this park. Neither predator can do much damage to an adult rhino, but calves, especially very young ones, are vulnerable. A lion's attack is swift, but unless a kill is made before the cow rhino can react, the lion has little hope of succeeding. Hyenas harass and nip at the youngster while dodging the charging adult. If there are enough hyenas in the pack to keep up the attack, death is a virtual certainty. Such kills have a major impact on rhino numbers in those parks where poaching has already greatly reduced the number of rhinos, but in Nairobi the adult rhino population and the number of young seem healthy enough to cope with the resident hyenas.

The giraffe population in the park is quite high and the animals appear to be everywhere. Nairobi National Park has one of the most studied populations of giraffes on the continent and many of the animals are known on sight by researchers. The giraffes found here are members of the Masai species and are easily identified by their irregular-shaped spots. No two individual giraffes share quite the same pattern, and it is these different markings that allow scientists to identify the individual animals.

From studies done in the park, it would appear that giraffes are not particularly good mothers. Young giraffes are rarely nursed after they are a month old and indeed may be separated from their mothers for long periods once past that age. Young giraffes have also been noted going to "kindergarten." It is not uncommon for mothers to leave their babies in a group while they go off feeding. For these reasons it is easy to understand why almost three out of every four giraffes born fail to survive their first year.

Giraffes are also known for their "necking" behaviour. Two males stand shoulder to shoulder and swing their necks at each other as a method of sizing up an opponent's strength and establishing a dominance hierarchy. Virtually every male giraffe knows who he can bully and who he can't. Most necking matches appear to be fairly friendly, but when a female in estrus is nearby, they can turn savage. Males have been known to be knocked over in such battles and some even lose consciousness. Even casual necking behaviour is an unusual sight on safari.

In the course of looking for rhinos or giraffes, herds of zebra, impala, eland and waterbuck are often encountered. Wildebeest, too, may be seen. There are around 500 wildebeest in the park. After a day spent in Nairobi National Park, it may seem like there is little else left to see in the grasslands. And in a sense that is true, for this park is a superb introduction to the ecology of the grassland. But it is rather like reading a *Reader's Digest* version of a good novel. It's all there but something is missing. That's the way it is with this park. It has all the players, the drama and the scenery, but it is a scaled-down version of the "real thing," the way it used to be.

Most safaris leave the next area for last, and for good reason. Take the total population figures for all of the animals found in Meru, Samburu, Buffalo Springs, Mount Kenya and Nairobi parks, and together they wouldn't equal the wildlife population to be found in the Masai-Mara. This area may well contain the identical types of animals found in Nairobi National Park, but there the similarity ends.

Opposite:
Male Masai ostriches develop bright red necks during the breeding season. He will have several females in his harem.

The female ostrich is very plain and drab. The females in the harem will lay their eggs in several different nests besides the one they will incubate. This behaviour helps insure the survival of at least some of her eggs even if her own nest is wiped out by predators.

Ostriches are often seen in the company of zebras and other plains game.

Opposite: This lioness managed to ambush and kill a male ostrich, a very rare event.
Overleaf: "Necking" among male giraffes helps establish a pecking order of dominance. The two on the left are not engaged in a serious battle.

The black rhino was once treated as vermin on the plains and shot on sight to help open the land to settlers. Today all rhinos in Kenya officially belong to the president and are protected and held in trust for the world.

Twenty-five black rhinos make their home in Nairobi National Park making this small reserve one of the best places in the country to see one.

A Cape buffalo with his head raised like this fellow's is an animal to be watched. He is sizing up the situation and he may retreat, stand his ground or charge!

Common waterbuck (m)

Bohor reedbuck

Common eland (f)

Lioness with cub

Lion cub

Serval

There are two species of baboon to be seen in Kenya. They are named by their colour; Yellow baboon and the Olive baboon. These are all Olive baboons. The male showing his teeth is performing an aggressive behaviour that displays his prime weapons to other baboons. The female, below, is grooming another baboon. Such "social grooming" helps bind the troup together.

Opposite: Rock hyraxs are the closest living relatives of the elephant.

A lone tree that somehow wound up in the middle of the Park's largest hippo pool provides a safe roost for some White spoonbills.

Hadada Ibis have one of Africa's best known sounds, a loud "har-har-har".

Opposite: These Ruppell's vultures roosting in the branches of a Fever tree are waiting for the sun to warm the ground. The warmed air creates thermals which the vultures ride into the heavens.

The Masai-Mara
The End of the World - Days 7 to 13

Masai-Mara Reserve
1,812 sq. km. (700 sq. mi.)
(518 sq. km. set aside along the lines of a national park, with no Masai cattle-grazing allowed.)

The jackal trots across the blackened earth, looking for an opportunity to feed. Yesterday the Maasai started a fire here to burn off the dead grass and to encourage new growth for their cattle. Such fires are a fact of life on the Mara and have been for over a hundred years.

The very existence of the grassland depends on the fires, for there is enough rain in the region to support a forest. The fire prevents the saplings from growing and helps ensure the continued bounty of grass for both cattle and wild animals.

Grasslands, no matter where they are found, support large herds of animals. The Mara supports the second-highest biomass (the total amount or weight of living things in a given area) of any land ecosystem. Only the world's rain forests outweigh it. But rain forests are so rich in variety that organisms have evolved into scores of species, each relying on some minute portion of the total for its livelihood. As a result, an individual tree might have living organisms on it that are found nowhere else in the world, and the forests in total house more different types of animals than anywhere else.

It is the grasslands, however, that support the largest herds. This is due to the abundance of one type of food that only a comparatively small number of species can use. Limited competition for that food results in huge herds. In North America that meant millions of bison; in Asia, millions of Saiga antelopes; and in Australia it was kangaroos.

But here in Africa's Serengeti - Mara ecosystem, the grassland reaches its zenith. Fueled by more energy from

Opposite: Blackbacked jackal, Thompson's gazelle, Masai-Mara.

the sun than any other grassland and blessed with sufficient rainfall, the great herds prosper.

All of this means little to the jackal on this particular morning. The fire is out and the sun, rising over the Loita Hills some kilometres distant, promises another good day.

The sun shines on a male Thompson gazelle that watches the passage of the jackal with casual interest. Most predators, once seen, cease to be much of a threat. The jackal isn't much of a threat to an adult Tommy in any case, but a predator is a predator and the buck doesn't take any chances.

Nighttime deprives the gazelle of its vision, and a short distance away there is ample proof of what can happen to the unwary.

The dead are not gazelles, but rather wildebeest. A herd of 50 or so wandered towards a small stream during the night, grunting as they went. Likely the wind was against their backs or they would have smelled the scent of the lions that lay in ambush ahead of them.

The lions, for their part, had come to the stream to drink. The grunting herd of wildebeest changed that, and the nine lion females in the group fanned out immediately. The dozen cubs stayed put.

The attack wasn't co-ordinated. Lions are not great hunters, as a rule, and miss on seven out of ten attempts at a kill. They don't think about the direction of the wind and often it spoils their hunt — but not this night!

The first lioness charged into the middle of the herd and brought down a yearling calf. The panicked herd split and ran into the other waiting lionesses on either side, and two more, a calf and a cow, were taken. Somehow a zebra was caught too, probably by the male lion, but its death was well away from the three other kills.

This morning little is left of the three wildebeest carcasses, except bones. The big, black-maned lion has confiscated the calf's head and busily licks out the brains while a few cubs and a lioness make a half-hearted attempt to get a tidbit. Not one of the 23 lions is left hungry, and most have gone to sleep.

Now the next players, the Cape buffalo, stumble onto the scene. This is a mixed herd of cows, calves and a few adult bulls. None of the calves are very young and all have lost their reddish-brown colour. It would be these half-grown buffalo that the lions would go after.

Were the buffalo just a little further away, the lions would show no interest at all. They are too full. But when the herd arrives practically in their laps, it is too good an opportunity to pass up.

Almost immediately the buffalo get wind of the lions and stand with heads held high, nostrils flared. There is no panic. They simply stand their ground.

The pride is certainly large enough to take even the biggest bull. For most of the year, when the wildebeest are absent, buffalo make up a substantial portion of their prey. The very size of the pride demands large animals to fill the many stomachs, and almost any buffalo provides more than enough meat.

Successful kills require an attack from behind. The buffalo's horns are too dangerous to risk a frontal attack. Today several sets of these lethal weapons face the pride.

What happens next may be some cleverly conceived strategy or perhaps just luck. The male and one of the females sit down right in front of the herd, almost ignoring them. This is too much for the buffalo and they charge. Immediately the lion takes off, followed by several members of the buffalo herd. He is careful not to get caught, but he appears to be equally careful to stay within sight, almost as if decoying the buffalo into a trap.

There is no question an ambush has been laid. With the buffalo distracted, the lionesses drop back and move into position, surrounding the milling buffalo. For a time it seems a large calf will be their next victim, but the herd closes rank and moves off. The lions are thwarted and watch as their quarry disappears.

Round one goes to the buffalo.

Elsewhere on the plain the grass is much shorter and provides no cover for lions, so the grazing herds have a measure of safety.

Because a predator can be spotted from some distance, the herds of gazelle, common zebra and wildebeest spread out, dotting the plain rather than clumped in tight groups as they were in the long grass.

Only certain types of predators can hope for much success here. The cats, with their smaller lungs, rely on ambush and are at a distinct disadvantage. Even the speedy cheetah has to get close before starting its lightning sprint, and the herds know enough not to let one get within range. Like other cats, the cheetah just can't take in enough air to sustain a long race.

Jackals, being dogs, have greater lung capacity, but they are too small. They are a threat only to the very young fawns, and today there are few babies to be seen.

A pack of 19 wild dogs, however, has no such restraints. Where lions succeed in only three of ten hunts, these predators succeed in seven out of ten. And this pack is hunting.

A few Masai giraffes watch the passage of the trotting dogs, but they are in no danger at all. The dogs simply ignore them and move on.

A distant herd of Tommies immediately become alert and begin moving off a good kilometre away. It seems the odds are against the dogs. They break into a run anyway, but soon give up.

A small herd of zebra appears to be the first realistic target, but they bunch and stand their ground. The dogs test them and find no weakness. Next they test a few wildebeest, but they too turn to face the dogs. The dogs keep running.

Kilometre after kilometre they run, gradually fanning out around a tree-covered hill, but no kill is made. The hill becomes a place to converge and regroup. Within a few minutes all of the dogs are present and all lay down.

After a few minutes more, one dog trots over and grabs a stick. Another tussles with him, and they look for all the world like pet dogs. Another pair nips and wrestles with each other, reinforcing the domestic scene. Then they all settle down. Are these the dreaded killers?

It looks like the day's hunt is over, but one dog rises and walks out onto the plain. She stands there staring out at the Thompson gazelles, which are sprinkled as far as the eye can see. Another dog stretches and walks up to her, then another and another. They sniff and greet each other, and then she starts out, her head low, her body stiff. The hunt is on again.

The gazelles respond immediately to the change in the dogs' posture and break into a run. Like a shotgun blast, the closely grouped dogs break formation and fan out, each after its own kill.

At first it appears as if this hunt too will be fruitless. The gazelles are too far, too fast. Still the dogs come.

A fawn is cut off from the rest of the herd. A dog veers after it. The fawn still has some distance, but the dog closes in on it. The fawn veers. The dog cuts in even closer.

Two jackals, then a third, join the chase, as if from

nowhere. The dog and the fawn ignore them. This is their race and theirs alone.

The dog literally bowls the youngster over, and the gazelle lays there, winded. The dog stands over his catch and looks back to see if other dogs are following. They aren't, and the jackals have stopped.

While the fawn watches, the dog bites, disembowelling it. Death is quick after that. The dog eats rapidly, gulping down chunks of meat.

The jackals edge closer, hoping for a chance to feed as well. Meanwhile two Ruppell's griffon vultures have observed the kill from the air and come to see what they might scavenge. Shortly after their arrival, a tawny eagle flies in, and it too waits for the dog to finish.

The dog keeps looking for the other members of the pack to join him, but they fail to appear, so he trots off to see where they are, giving the scavengers their chance. They rush in. The dog hurries back to chase off the intruders. Soon it becomes apparent that no other dogs are going to show up, and reluctantly the dog leaves. Most of the remaining meat goes to the jackals.

The reason no other wild dogs showed up at the fawn kill was because the others succeeded in killing two more Tommies, an adult female and a second fawn. Like the first, they were eaten alive and then torn to pieces as dog after dog rushed in to get its share.

With such a high success rate when hunting, why aren't there more wild dogs on the plains? There are two major reasons.

The first is a product of the dog's social system. If disease strikes one dog, it will almost certainly strike the entire pack. Rabies, distemper and other canine diseases ensure that the packs never get too out of hand.

The second reason is that, for the most part, man's hand is against them. All wild canines — wolf, coyote and wild dog — kill large prey by literally bleeding the prey to death. Often they will disembowel their prey and feed on it while it lies helplessly watching. By human standards it is an ugly, savage death. In contrast, when a lion kills, it kills by clamping its jaws over the prey's mouth or throat and suffocating it, or by breaking the animal's neck. Either way, it is less gory and apparently more acceptable to humans.

As a result wild dogs have been hunted as vermin across much of their range, resulting in a steep decline in their numbers. On the other hand, they are also much respected for their loving and caring role as nature's ideal parents. Perhaps neither viewpoint is correct. The dogs, like the lions, are just one important cog in the grand scheme of the grasslands.

* * * * *

Death may seem like a common occurrence on the plain, but that impression is far from the truth.

The most impressive thing about the Serengeti - Masai-Mara ecosystem is the abundance of life to be found there. There are over two and a half million large mammals. Of that figure, less than 5,000, perhaps only 3,000, are lions. There are fewer than 100 wild dogs most years, and at times perhaps only 20. There are about 5,000 hyena, and maybe a few hundred leopards and cheetahs. The chance of being killed by a predator, given those odds, are slim on any given day for any single animal.

Giraffe have little to fear from wild dogs but it never hurts to check out any predators in the area.

A lion is well equipped for its role as the dominant predator of the plains.

The order of dining at the lion's table is: male first, female second and cub last. It is most often the lion and not the lioness that is likely to allow cubs to feed on a kill if food is scarce.

Sometimes food is abundant and everyone feeds and then as a group the pride seeks whatever shade is available and sleeps the day away.

Sometimes rest period is disturbed when "company" blunders in. In this case the "company" is a herd of Cape buffalo.

As this sequence shows the herd of buffalo spotted the large male and chased him off. The lion ran in a semi-circle and led the herd back into the waiting lioness but they were spotted too and the whole incident ended in a stand-off.

Thompson gazelles are one of the most common grazers on the plain.

They are a favourite prey of the wild dogs, one of the plain's most successful hunters.

After resting, one of the dogs, a female, stood and gazed out onto the plain . . .

Thompson gazelles were everywhere and some of them were very young and vulnerable.

The hunting posture of dogs usually alerts the gazelles to danger some distance away . . .

but this time they let the pack get too close.

The dog knocked down the Tommy and then proceeded to eat it alive.

The kill attracted many scavengers including a jackal.

Normally, a dog shares its kill with other members of the pack but this time the others had made their own kills out of sight and left the lone animal to contend with the jackal as best it could.

The arrival of vultures proved too much for it and the dog left leaving the scavengers to fight over the remains.

It is a herbivore's paradise.

Each plant-eater has its own niche, so rarely do two species compete for the same food. In the wooded areas, Masai giraffe browse on the tops of trees while herds of impala feed on the lower branches.

Even in the grassland, an area which at first glance appears to consist of only one type of plant, there is a variety of food to select from. Zebra prefer the coarser grasses, while topi, with their narrow noses, select the finer new shoots. Thompson gazelles prefer another type of grass, while the slightly larger Grant's gazelle selects still another variety. It may appear that several species are feeding on the same plant when they are seen together, but that is an illusion.

The elephant is one animal that does compete with several species. It eats both trees and grasses, and in some areas it can have a detrimental effect on local herbivores. In the past this was less of a problem because the elephants moved constantly over their range, never staying long in an area. That freedom is more and more being denied to the massive beasts, and as their range is restricted, the ecological damage they do may one day upset the balance, causing other species to disappear. Scientists are rightly concerned about this possibility and several long-term elephant studies are under way to learn more about the elephant's role in the ecosystem.

Despite the increasing numbers of elephants in the region, there has been little noticeable impact as yet on other species. In fact, following massive slaughters of game after World War II, the herds have been steadily increasing. For instance, in 1959 the population of Grant's and Thompson gazelles was only about 200,000 animals. By the mid-70s the Tommies alone numbered 500,000. Zebra increased in the same period from 60,000 to 200,000, buffalo from about 2,000 animals to 50,000.

Outnumbering all of these, though, is the wildebeest, and they are arriving on the Mara's plains in ever-increasing numbers. In the late 1950s there were perhaps 250,000 wildebeest in the ecosystem. Today their numbers hover between one and two million animals. And while other species have stabilized, the wildebeest herds may still be increasing.

This Serengeti - Masai-Mara herd is the largest such congregation of free-roaming mammals remaining on Earth. Each summer the herd leaves the southern calving ground in Tanzania's Serengeti National Park and moves north to where the rains have brought fresh growth. For the next nine months they roam, covering about 500 kilometres on their migration before returning to give birth. Sometime in July a trickle of these animals begins to cross the Kenya - Tanzania border, mixing with the Mara's resident herds of wildebeest or gnu.

By August that trickle has become a flood and the plain is covered with wildebeest. By late fall they will be gone, but for now they are everywhere.

The arrival of a few million animals tends to have some impact on the environment. Lions and hyenas benefit from an enriched larder, while fellow herbivores may find themselves eaten out of house and home.

Despite the competition for food, other plant-eaters mix with the passing herds of wildebeest. This year the migration is sporadic because there has been plenty of rain to nourish the plains. Given no clear destination, the various herds of wildebeest haven't meshed but drift independently in groups numbering from less than 100 to over 5,000.

Small herds of zebra can often be found with the gnu. The large herds of zebra, however, generally precede the wildebeest and are already further north in the longer grass.

Even with the rich supply of food, the herds of wildebeest move on, pushed forward by increasing numbers of their own kind. En route they must climb a small hill. It is no obstacle except that, being higher, it has a slightly enriched climate that supports some tree growth and underbrush. The hill is called Rhino Ridge, and feeding on the bushes is its namesake, a female rhino and her calf — the last two of their kind left in the Mara.

The wildebeest pay her no heed and continue on. There is another animal hidden here too, and unlike the rhino it does present a threat to the wildebeest. This time a cow wildebeest is killed by three young adult lions.

The lions quickly gut their kill and one of the two females drags it into the shade of a large tree. The bloody remains immediately attract several vultures. Vultures are quite rightly known as Africa's sanitation crews. By cleaning up the remains of dead animals, they help reduce the spread of disease and thus help keep the herds healthy.

The first to arrive at the pile of intestines is a lappet-faced vulture. It is the largest of the vultures and has a wing span approaching two metres (8 feet). Each of Africa's six species of vulture has developed a specialized method for feeding on a carcass. The lappet-faced, for example, has a strong beak that can open a carcass up, and so it is often the first to feed. It is also one of the few vulture species that occasionally kills its own prey. It is not a social bird and is usually seen alone or in twos.

Two other, smaller species soon arrive and begin fighting over the remains. About equal in size, both the Ruppell's griffon vulture and the less numerous white-backed vulture are equipped to reach well into a carcass after the lappet-face has opened it. As a way of reducing the chance of contamination, their heads lack feathers so that the gore may easily be cleaned off. Today that will not be much of a problem because the three lions have no intention of giving the dead wildebeest to them. They must be content with what little is left behind.

A tawny eagle and a Marabou stork also join in the fray, but the vultures leave little and soon all five species retire to nearby trees.

Not too far away a herd of long-faced kongonis are watching a bit of comic relief unfold. A warthog has ventured into a pack of six hyenas. Warthogs can be easy prey if they run, but this one stands its ground.

This is its lucky day, for only one of the hyenas is willing to attempt a kill. It approaches the warthog fully expecting to see it race off. When it doesn't the hyena must consider the length of tusks the male pig is sporting. Are they worth the risk? With no help coming from the other five hyenas, the would-be killer backs off and returns to a wildebeest head left over from last-night's hunt.

Spotted hyenas were once thought to be cowardly scavengers. It is now known that they are better hunters than lions, and indeed it is the lion who scavenges more often from the hyena and not the other way around.

Hyenas hunt like wild dogs. They can run for miles after their prey, if need be, and like the dogs, they too kill by biting out chunks of their prey.

Unlike dogs, hyenas can and do feed on larger herbivores, and in doing so compete for the same food supply as lions. The population of both hyenas and lions is about equal, and scientists wonder which of the two predators will survive in the long run as they fight to occupy the same niche. Significantly, lions will kill but not eat hyenas.

From Rhino Ridge it is less than a half-day's walk for the wildebeest to the Mara River. En route they will pass through the territory of another large pride of lions, but today they will do so safely.

The Mara is not a big river by any standard, but it is the biggest in the park. For some reason the wildebeest feel compelled to cross it. It may be that there is better grass on the far side, or perhaps the pressure from the herds behind them builds up, forcing them on, but for whatever reason, they must cross it.

The first obstacle is the thick brush lining the river. Are there predators waiting there? The herd which was spread out bunches and, after several false starts, begins to move cautiously forward. A herd of zebra has joined them.

There is no evidence of lions and the herd quickly reaches the upper bank of the river. Here the leaders stop and survey the crossing place.

Hippos are very much in evidence in the pools above and below the crossing place. Despite their large tusks, hippos are grazers and leave the safety of the river each night to wander the plains and feed on grass. Their rutted trails to and from the river are the paths the wildebeest will have to use to scale the steep bank that faces them. Without the hippo trails, the crossing would be all but impossible.

A more immediate problem confronts the mixed herd, for lying on the near side of the river are several crocodiles. There are more up- and downstream too.

In order to enter the river, the wildebeest must go down to the sand bar that the closest reptiles are occupying.

Typically, the wildebeest panic and race back. But five minutes later the grunting mass returns, and this time the herd moves right down, almost nose to nose with the crocs. There, for a few seconds, they stop and stare at their enemy. It is a brief truce because the lead animals are suddenly jostled forward by the force of those behind them, and given no choice they plunge into the water. The crocs beat them by only a few seconds.

What follows is a scene of absolute confusion. Wildebeest are good swimmers and quickly cross the river. A problem far deadlier than either crocodiles or lions awaits them at the other side. The bank which allowed the first herds here to cross easily has been eroded and worn down by the passage of so many wilde-beest, and the hippo trails have almost completely vanished. There is virtually no way up. The first few wildebeest find the only passable spot and, along with a few zebras, manage to scale it. Behind them are hundreds of other animals trying to squeeze into the same space. They bunch and panic. One tries to leap up and falls back

Opposite: Elephants are increasing on the Mara due to increased pressure from poachers outside the park.

on the others. Then another tries and falls. The few zebras in this melee quickly turn back. They will cross another day. But the wildebeest stay.

Soon the weaker ones are pushed under and trampled. Bodies begin to drift downstream, where they catch on the protruding legs of other dead wildebeest left over from an earlier crossing. The madness goes on, and then some wildebeest turn back, swimming past others just entering the water.

The crocodiles attack with deadly effect. First one wildebeest and then another jerk their heads up as the reptiles' jaws clamp on their bodies. For the most part, the killers are never seen by their victims. The crocodiles' green bodies are hidden by the muddy Mara River, and soon so are the bodies of the half-dozen gnu that have also been killed.

Half-grown calves can't quite manage the far bank, and they are the group that suffers the greatest losses. Most do make it safely back to the bank they started from and bleat out a call for their mothers. The mothers are calling for them too, but from the top of the bank on the other side. Eventually they turn their backs on their babies and join the thousands of other wildebeest already across. The calves continue to bleat.

Toward the end of the crossing, one calf turns back to join the other youngsters. Today is its lucky day! It is in chest-deep water when the huge jaws emerge from the Mara. But the crocodile has misjudged this time and it can't get a grip on the wildebeest's body. The calf escapes and is trotting up to rejoin the others that failed to cross when the lioness attacks. Despite all it has been through, the calf still has enough strength left to outrun her, and escapes.

The lioness stops and watches as the panicked herd departs. The calf has survived a harrowing day, but tomorrow it will return along with the next herd. The steep bank will still be there and so will the hippos, crocodiles and the lioness.

The wildebeest that died on this crossing probably numbered around 30. It must be remembered, however, that out of the hundreds of thousands on the plain, 30 is a mere handful. Being the most successful animal on the plain has a price, for it also means that the species will be the one to suffer the greatest losses.

A quarter of a million calves will be born next spring. Of those, half will die within hours of birth. Those that survive those first few hours will do so because the wildebeest all give birth within a few short weeks and the plains are so rich with easy pickings that the predators can't possibly kill them all. Of the remaining calves, about half of them will die during the migration, but the 30-or-so percent that do live will replace the adult losses and swell the herd's numbers.

Meanwhile, some distance away, the big pride of lions has moved back to the forested ridge where they like to rest. The 23 cats sprawl in the long grass, doing what lions do most of the day; they sleep.

One of the two large males that protects the pride's territory has moved slightly off from the rest, along with a lioness in heat. Lions mate several times a day, and this pair is no exception, but now they too sleep.

The other male, probably a brother to the first, has been away inspecting the boundaries, and upon his return he sniffs a spot where one of the other females has recently urinated. In the odour, his nose detects the distinctive scent that tells him that she is in breeding condition, and he performs a flehem. This look is often mistaken for a lion roaring, but in fact no sound accompanies the grimace.

Although undoubtedly excited, he will not challenge his brother and will wait his turn to breed. For now, he too will sleep.

Scattered across the plain are a sparse population of cheetahs. While a true cat, they differ from lions, leopards and others in one particular way. They lack retractable claws. This deprives them of the deadly tearing capability of the bigger felines, but that is of little consequence. Cheetahs do not need claws, for they kill primarily by strangulation.

Also, cheetahs rely more on their speed, which is unequalled among land-dwellers, than on stealth. A prey animal tripped up by a cheetah at speeds in excess of 100 kilometres an hour may well break its own neck in the tumble.

Such specialization has refined the cheetah into a superb predator. It has also doomed it. Cheetahs are over-evolved and probably lack the genetic capability to respond to environmental changes. There is so little difference between one cheetah and the next that scientists speculate that its biological clock has almost run out. Its extinction may not happen for a hundred years, or even a thousand, but in geological time its days are numbered.

It is arguably the most beautiful of cats, and the Serengeti - Mara ecosystem is home to one of the largest remaining populations. They are usually seen alone or in family groups consisting of a female and her offspring.

Near the lions' ridge, the cheetah population is represented by a single, very young cub and its mother. The mother's attempts at hunting are more often than not frustrated by the cub's overabundance of enthusiasm.

On the edge of the short grass plain where the wild dogs hunted is another female with three nearly grown cubs. Near the herds of wildebeest are two more, a courting pair, who have little hope of killing even the nearly grown calves but instead concentrate on smaller game.

And then, against all logic, there is the odd and inept group of five. Four females, one male, all existing in a small cohesive pack. What makes them odd is that four of them are related, but one, a female, just sort of appeared.

Normally rangers don't interfere in the natural course of life here, but with these cheetahs, that policy was bent if not outright broken.

Several months ago the mother of the four cubs tried to take a warthog. She was gored, and although she survived, her great speed was gone. Because of the rarity of cheetahs, the park decided to feed the cubs to help out until the adult recovered. A hyena caught her and ended any such hope. Orphaned, the cubs looked to the rangers for food and were eventually joined by another young freeloader.

Now nearly grown, they are almost weaned of park assistance, but they still lack enough experience to be good at killing their own food. One begins a hunt only to have the four others start a game of tag. End of hunt.

That was more the rule than the exception, but they do succeed in killing occasionally, usually when luck intervenes on their behalf.

This late afternoon, circumstances conspire to present them with dinner, a male impala. Impala have evolved a social system that is different from the other herbivores on the plain. They are not nomadic and are year-round residents of the wooded areas of the Mara, where they exist in three distinct types of groups.

The first group consists of females and young. They move about as a herd, and at times they move into an area occupied by the second group. These are the territorial males. The entire wooded area is divided into territories, each occupied by a dominant adult male. When the females wander into a territory, the buck does everything in his power to keep them there, for at least one and perhaps several females will be receptive to his advances. The browsing, however, is limited and the female herd may soon move on into the next male's territory.

Along the edge of the territories may be found the bachelor herds, consisting of young and adult males. Here the nonterritorial males can rest and hone up the skills needed to displace a territorial male.

It is one of these males that challenges a territorial male not far from where the five cheetahs are resting. The territorial male is tired from defending his harem against other males and from the demands of fathering future generations. He hasn't eaten much since the females arrived and so is doubly vulnerable to a challenge.

The two adult males fight, but it is brief, and the former ruler is ousted quickly. He runs off, chased by the new dominant male. The new harem master doesn't go far before returning to the females, but the loser keeps running.

And he runs right into the five cheetahs. The chase is brief but furious. The end not at all inevitable, but time and circumstances have decided in favour of the cheetahs. Death, at the hands of this inept group, is not quick by cheetah standards, but finally they succeed.

The cheetahs quickly open up the carcass and gulp their food. Everyone steals from cheetahs — lions, hyenas and even large flocks of vultures — so the slim cats can't afford to eat slowly.

They constantly look around for any other predators, but this time none show up. One reason is that the vultures, which usually are the first to arrive at a kill, are now roosting. Without the vultures to guide them, the other predators don't know about the kill. Luck again is a factor.

Meanwhile, back at the lion pride, the larger male and his female have wandered off into the woods for a bit of privacy. Romance, however, is quickly forgotten when a herd of buffalo stumbles between them and the rest of their pride. Will there be a hunt? For a few seconds it looks like the female might attempt an attack, but then the buffalo charge, sending the lovebirds scurrying for cover. There is no decoying attempt this time, just outright panic.

In the end the king of beasts and all 22 of his pride are crowded on a rock or hiding in the thickest bush while the buffalo herd drifts peacefully on.

Over the escarpment which defines the western boundary, the sun is setting. The heat of the day has long gone and the night will be chilly. Zebras and gnu will continue to feed. The Thompson and Grant's gazelles will bed down, and the hippos will emerge from the Mara to feed. Somewhere a hyena chuckles, and out on the plain a lion roars.

Topi

Cape buffalo bull

Thompson's gazelles and common zebra

Grant's gazelles sparring in front of a resting Tommy

*Opposite: Zebra are soon followed onto
the Mara by huge herds of wildebeest.*

Wildebeest don't like long grass and tend to move through it rapidly in single file.

They have good reason for their concern for this is an ideal place for an ambush.

A single lioness is quite capable of killing any unlucky wildebeest she might catch. This one was helped by another female and young male.

The Lappet-faced vulture is the largest of Kenya's vultures.

When this warthog wandered into a pack of resting hyena one of them started a hunt. Either the warthog's long tusks or the fact that the hyenas had killed a wildebeest earlier that morning made pursuing the matter very unattractive.

Opposite: Wildebeest hurrying towards the Mara River.

At the river the wildebeest encountered an animal most of them had probably never seen before.

Crocodiles and hippos are very common along the river but only the croc presents a real danger to the wildebeest.

The badly eroded bank provided few places for the animals to escape the river.

Still more and more animals leaped into the river adding to the confusion.

The zebras were the first to realize the danger and crossed back to safety.

It was most difficult on the calves for they weren't as big or as strong as the adults.

The animals raced along the edge of the river looking for a way up.

Soon animals were crossing and re-crossing the Mara.

This is what lions spend most of their day doing, about 18 hours on average.

A male lion can all but disappear in the long grass if he chooses.

The cheetah is usually a solitary cat unless it has young or is courting.

Young cheetahs have a mane that gradually disappears as they mature.

An impala buck scent marks a bush.

The impala's leaps serve to confuse
a predator.

The fastest land mammal alive today!

Cheetahs are nervous feeders. They are always on the lookout for lions or hyenas that might steal their kill.

Tawny eagle

Secretary bird

Saddlebilled stork

Yellowbilled stork

The Dry Lake - Day 14

Amboseli National Park
3,810 sq. km. (1,171 sq. mi.)

Amboseli is a land of stark images that evoke the very essence of Africa and the magic of safari. It is also a land of contrasts.

It is a flat land. There is only one hill within the entire park, and that not an impressive one. Yet dominating the entire area, even on a heavily overcast day such as this one, is Africa's highest mountain, Mount Kilimanjaro. This ancient volcano, 5,894 metres (19,332 feet) high emerges just south of the Kenya - Tanzania border and forms a perfect backdrop for the park's drama.

At this time of year it is a land of drought. Earlier in the year Lake Amboseli floods the plain, but by August all traces of water have vanished. Left behind is the white, chalky bottom across which wildebeest and dust devils track their ways. But there *is* water, and plenty of it, in two year-round swamps at the east end of the park.

The water that feeds these swamps comes not so much from Kilimanjaro's snowfields as from distant hills. The precious moisture trickles underground only to emerge here in springs which enrich the dry land. For a short time it is a river, but soon it once again disappears beneath the ground outside the park. Eventually some of that water bubbles up again in Mzima Springs in Tsavo National Park before completing its journey to the Indian Ocean.

On the dry plains dust swirls around the bleached bones of hundreds of animals. It looks as if a massacre has occurred, but there is no more death here than in any other park. Here, without the grasses and bushes to obscure them, the bleached bones stand out as bleak reminders of the daily struggle to survive.

Scattered among the bones are signs of animals. Droppings abound, as do scores of tracks. All these signs

Opposite: African fish eagle, Amboseli.

attest to the abundance of wildlife here.

Lions court and mate while buffalo plod by. Cheetahs seek shade under one of the trees in the park's clumps of bushland. A hyena rests at a water hole, quietly watching the comings and goings of its prey.

On the plain, a Masai giraffe gallops, sending the ever-present oxpeckers flying as they lose their balance. A lone wildebeest treks across a barren landscape while a large herd gathers in the green grass around the swamp to feed.

The garish barking of a zebra stallion breaks the silence as it chases after its small harem, while another races through the swamp.

A huge one-tusked male elephant moves majestically through the brush while a cow herd shuffles toward the long elephant grass on the far side of the dry lake bed.

It is all here, the big game that Africa and Kenya are famous for, along with the backgrounds that have become familiar from countless postcards, books and movies.

Often overshadowed by these spectacular sights are the other, smaller residents — by no means less spectacular — the birds of Amboseli.

A birdwatcher can find suitable subjects in any of the parks. Kenya boasts about 1,500 different resident and migrant species of birds, of which some 400 are to be found in Amboseli. There are 47 different species of birds of prey alone recorded here. The swamp tends to concentrate them, and a few hours spent around its border is bound to produce a variety of species.

Pink-backed pelicans are more white than pink, and they can be found fishing the open water of the swamp. Other fish-eaters are here too. Members of the heron family are common. Among the reeds it is not unusual to encounter both grey herons and the giant Goliath heron stealthily stalking their prey. A third large heron, the black-headed heron, is less fond of fish and instead stalks lizards in the tall grass. The cattle egret, another member of the heron family, prefers insects stirred up by the passing of large game and can often be found near these herds.

Some of the birds of prey are equally common near the swamp. Eagle owls and fish eagles share similar niches, but one prefers to hunt by night, the other by day.

Marabou storks are occasionally encountered in large groups along the shore. While usually associated with carrion, they can and do catch small fish and reptiles when the opportunity presents itself. Marabous can be quite deadly to newly hatched crocodiles if their mother is absent from the nest, but here in Amboseli there are no crocodiles at all. There are, however, skinks, agama lizards and a host of small reptiles that fall easy victims to this stork's powerful bill.

Marabous appear to be essentially lazy birds and seldom appear to stalk their prey with much effort. The Kori bustard, a large grey bird, does stalk the plain for prey but is equally at home eating grass seeds.

Another seed-eater is the chicken-like vulturine guinea fowl. At one time their flocks numbered in the thousands, but habitat destruction and hunting have greatly reduced their numbers. Flocks of up to a hundred are frequently encountered in the park, however.

Other birds common around the fringes of the swamp include blacksmith plovers, the sacred ibis, stilts, African darters, and others that appear and then dart away before their identity can be ascertained. The largest flocks are almost certainly made up of sand grouse, which arrive by the hundreds to quickly drink and fly off.

Songbirds are equally numerous and come in a variety of sizes and shapes. There are several species of weaver birds to be seen, each with their distinctive nests. Sometimes among them may be found a variety of starlings, including the beautiful Superb starling, a bird which considerably outshines its more northerly namesakes.

Flitting around the branches may be seen a bird found only on the African continent. The mousebird appears to have its feet located on the wrong part of its anatomy and this trait, combined with its long tail, reminded settlers of mice (for reasons known only to them). Hence the name.

Two species of oxpeckers, sometimes called sentry birds, may be seen pecking a variety of ticks off of the backs of buffalo and other hosts. They are called sentry birds because they are believed to warn the host animal of approaching danger.

Ostriches too are found here, but are less numerous than in other parks. There is another two-legged creature walking the park and it is one that should not be here according to Kenya's park rules. When Amboseli National Park was created out of what was originally a huge game reserve, the Maasai tribe was given certain trade-offs, in return for which they agreed to keep their cattle out of the core area of the park. They were to use the vast game reserve surrounding the park for grazing and for pursuing their traditional lifestyles. But this August they and their cattle are everywhere in the park, and their presence highlights the classic dilemma of emerging African nations.

In the Masai-Mara Game Reserve, the core of which functions exactly like a national park, the same rules govern the comings and goings of the Maasai and their cattle. They keep their livestock outside the area set aside for wildlife, but in the Mara there is more rain and better grass than in the Amboseli region. This particular year the area is in a severe drought and traditional water holes outside the park have dried up. With nowhere else to turn, the herds of goats and cattle have been driven into the park, where there is year-round water — a direct contravention of the park's rules.

There is little than can be done to stop them, and perhaps in times of stress no action should be taken. Somehow Kenya and other conservation-minded nations must grapple with this classic problem: the needs of people versus the needs of wildlife.

In fairness to the Maasai, it should be pointed out that, given their pastoral lifestyle, they really are responsible for many of Kenya's richest wildlife habitats remaining rich. They have no need to hunt and they approach wildlife with a live and let live attitude that helps preserve the great herds. But it seems no problem concerning wildlife can be dealt with in simple terms. In the old days the Maasai, in order to prove their manhood, hunted lions and other dangerous game with spears. This practice has since been outlawed, but in fact it is still practised from time to time. Elephant researchers in Amboseli report finding speared elephants on a regular basis.

A few such losses are a minor problem, but as the Maasai population turns away from the nomadic lifestyle of the past to embrace farming and townlife, attitudes change. In this they are neither better nor worse than any other group that finds itself confronted with poor wages and destructive wildlife. Poaching results. In a country where a single rhino horn equals a year's wages on the black market, it is not difficult to understand why there are no more rhinos. Amboseli, once world famous for its long-horned rhinos, has virtually none left. They have been poached.

In a country that hopes to see one million tourists a year, wildlife represents a major draw. A single rhino, alive, over the years will bring into the economy much more than its horn will. Kenya and her people must come to terms with the consequences of short-term versus long-term gains.

It has not been pointed out, especially by North American writers, but Kenya's wildlife preservation track record — along with the records of many other African nations — is better than ours. Critics of African game management techniques say that soon all of the elephants will be confined to national parks or reserves, but they overlook the fact that North America's biggest game animal, the bison, exists almost entirely in such parks now.

South of Canada there are virtually no free-roaming grizzlies, but throughout Kenya there are free-roaming lions, leopards, elephants and other big game.

Kenya's track record, to this writer's mind, is better than ours. They may still be able to avoid the pitfalls that conservationists in North America fell into in the 19th and early 20th centuries. Their country is still rich and bountiful.

Despite a new awareness of the many concerns facing Africa today, I ended my journey to the place where the grass and the sky meet, that place the Maasai call The End, with the feeling I had been to the very beginning of time. The world owes Kenya and other like-minded nations a tremendous debt.

Cattle egrets often follow large animals feeding on the insects they stir up.

A lioness can be quite coy when she is being courted.

Common zebra

Overleaf: Around midday the dust devils begin to appear all over the dry lake bed.

Maasai water their cattle in the park in times of severe drought.

A Yellowbilled oxpecker seeks the insects and ticks that live on this Cape buffalo bull.

Hyenas love to cool off in the swamp at midday.

Cape buffalo escape the flies and the heat in the water too.

The Goliath heron is the largest in Kenya.

Pink-backed pelicans

Grey heron

Marabou storks

Blacksmith plover

Kori bustard

Sacred ibis

Lilac-breasted roller

Superb starling

Ground hornbill

Black-necked heron

Bibliography

Adamson, George. *My Pride and Joy*. London: Collins Harvill, 1986.

Adamson, Joy. *Born Free*. London: Collins Harvill, 1960.

Beard, Peter. *The End of the Game*. New York: Viking Press, 1977.

Cott, Hugh B. *Looking at Animals*. London: Collins, 1975.

Douglas-Hamilton, Iain & Oria. *Among the Elephants*. London: Collins Harvill, 1975.

Fetner, P. Jay. *The African Safari: The Ultimate Wildlife and Photographic Adventure*. New York: St. Martin's Press, 1987.

Fletcher, Collin. *The Winds of Mars*. New York: Alfred A. Knopf, 1973.

Grzimek, Dr. Bernard. *The Serengeti Shall Not Die*. London: Collins, 1960.

Insight Guides. *Kenya*. New York: Prentice Hall Press, 1988.

Iwago, Mitsuaki. *Serengeti*. San Francisco: Chronicle Books, 1986.

Kruuk, Hans. *The Spotted Hyena*. Chicago: The University of Chicago Press, 1972.

Matthiessen, Peter, and Eliot Porter. *The Tree Where Man Was Born/The African Experience*. New York: Avon, 1974.

Moss, Cynthia. *Elephant Memories: Thirteen Years in the Life of an Elephant Family*. New York: Wm. Morrow and Co., 1988.

Moss, Cynthia. *Portraits in the Wild: Behaviour Studies of East African Mammals*. Boston: Houghton Mifflin Company, 1975.

Myers, Norman. *The Long African Day*. New York: The Macmillan Company, 1972.

Reader, John, and Harvey Croze. *Pyramids of Life*. New York: J.P. Lippincott Company.

Russell, Franklin. *Season on the Plain*. Toronto: McClelland & Stewart, 1974.

Scott, Jonathan, and Brian Jackman. *The Marsh Lions*. London: Elm Tree Books, 1982.

Schaller, George B. *Golden Shadows, Flying Hooves*. New York: Alfred A. Knopf, 1973.

Schaller, George B. *Serengeti: A Kingdom of Predators*. New York: Alfred A. Knopf, 1972.

Schaller, George B. *The Serengeti Lion: A Study of Predator-Prey Relationship*. Chicago: The University of Chicago Press, 1972.

Van Lawick, Hugo. *Among Predators and Prey*. London: Elm Tree Books, 1986.

Van Lawick, Hugo. *Savage Paradise: The Predators of the Serengeti*. London: Collins, 1977.

Van Lawick-Goodall, Hugo and Jane. *Innocent Killers*. London: Collins, 1970.

Williams, J.G. *A Field Guide to the National Parks*. London: Wm. Collins, 1986.

Zeisler, Gunter. *Safari, The East African Diaries of a Wildlife Photographer*. New York: Facts on File, 1984.

WORKBOX

This chapter gives you all the information you'll need to produce perfect cross stitch embroidery and successfully recreate the projects in this book. There is advice on materials and equipment, embroidery techniques, stitches and making up methods.

MATERIALS AND EQUIPMENT

The materials and equipment needed for successful cross stitch are minimal. This section describes the basics you will need to complete the projects in this book.

Fabrics

Most of the designs in this book have been worked on Aida fabric which is stitched over *one* block. In the main, the size used is 14 blocks or threads to one inch (2.5cm), often called 14 count. Some designs use an evenweave fabric such as linen which should be worked over *two* threads. The same design stitched on fabrics of different counts will work up as different sizes. The larger the count (e.g. 18 count), the more threads per inch (2.5cm), therefore the smaller the finished design, and vice versa. Each project lists the type of fabric used, giving the thread count and fabric name, which should be quoted when purchasing goods. All DMC threads and fabrics are available from good needlework shops (see also Suppliers).

Threads

If you want your designs to look exactly the same as those shown in the photographs, you need to use the colours and threads listed for each project. The threads used in this book are DMC stranded cotton (floss). On some of the projects I have suggested that they could be stitched with tapestry wool (yarn) instead.

It is advisable to keep threads tidy and manageable and thread organisers and project cards are ideal for this purpose. Cut the threads to equal lengths and loop them into project cards with the thread shade code and colour key symbol written at the side. This will prevent threads from becoming tangled and the shade codes being lost.

STRANDED COTTON (FLOSS) This is the most widely used embroidery thread and is available in hundreds of colour shades, including silver and gold metallic. It is made from six strands twisted together to form a thick thread, which can be used whole or split into thinner strands. The type of fabric used will determine how many strands of thread you use: most of the designs in this book use two strands for cross stitch and one strand for backstitch.

TAPESTRY WOOL (YARN) DMC wool is a matt, hairy yarn made from 100 per cent wool. It is made from short fibres twisted together to make a thick single thread which cannot be split. Designs using tapestry wool are usually worked on a canvas using one or two strands. A wide selection of colours is available, with shades tending to be slightly duller than for stranded cotton. There are conversion lists for colour matching from stranded cotton to tapestry yarn – ask at your needlework shop.

Needles

Stitch your designs using a tapestry needle which has a large eye and a blunt end to prevent damage to the fabric. Choose a size of needle that will slide easily through the holes of the fabric without distorting or enlarging them. If using beads to enhance a design you will need to use a beading needle, which is thinner and longer. You will probably find it easier to sew if you use a thimble, especially for canvas.

Scissors

You will need a sharp pair of embroidery scissors for cutting your embroidery threads and also a pair of good dressmaking scissors for cutting fabric.

Embroidery Frames

Your work will be easier to handle and stitches will be kept flat and smooth if you mount your fabric on to an embroidery hoop or frame, which will accommodate the whole design. Bind the outer ring of an embroidery hoop with a white bias tape to prevent it from marking the fabric. This will also keep the fabric taut and prevent it from slipping whilst you are working.

BASIC TECHNIQUES

The following techniques and tips will help you attain a professional finish by showing you how to prepare for work, work the stitches and care for your finished embroidery.

Preparing the Fabric

Spending a little time preparing your embroidery fabric for work is a good idea, helping to avoid mistakes and produce superior finished results.

FABRIC SIZES Start by making sure you are using the correct size of fabric. You can check this by looking at the stitch count (the number of stitches across the height and width of the design) and design size given with each project. Each project gives the finished size of a design when worked on the recommended fabric, together with the amount of fabric needed. The overall fabric size should be at least 8–10cm (3–4in) larger than the finished size of the design to allow for turnings or seam allowances when

mounting the work or making it up. To prevent fabric from fraying, machine stitch around the edges or bind with tape.

Measurements are given in metric with the imperial equivalent in brackets. Always use either metric or imperial – do not try to mix the two.

CENTRE POINT Starting your stitching from the centre point on the fabric ensures you will have enough fabric all round the design. To find the centre point, tack (baste) a row of stitches horizontally and vertically from the centre of each side of the fabric. These lines correspond to the arrows at the side of each chart and will cross at the centre point.

Using Charts and Keys

All the designs in this book use DMC embroidery fabrics and stranded cotton (floss). The colours and symbols shown on the chart key correspond to DMC shade codes. Each project lists the number of the skeins required for each thread colour together with a colour name, which is given for easy reference only – when purchasing threads, use the code numbers.

Each coloured square on the chart represents one complete cross stitch and some squares will also have a symbol. The colours and symbols correspond to those in the key at the side of each chart. A triangle in half a square represents a three-quarter cross stitch. French knots are indicated by a coloured dot – the project instructions specify what thread shade to use. The optional use of beads on some designs will be in the instructions and will also specify which colours they replace. Solid coloured lines indicate backstitch or long stitch – refer to the project instructions for details.

Small black arrows at the side of a chart indicate the centre, and by lining these up you will find the centre point. Some of the charts are spread over four pages with the colour key repeated on each double page. Small blue arrows on some of the charts indicate the cut-off line for the use of that design on another suggested project.

To prevent mistakes, work systematically so that you read the chart accurately. Constantly check your progress against the chart and count the stitches as you go. If your sight is poor you may find it helpful to enlarge a chart on a colour photocopier.

Adding Beads

Replacing a single thread colour with beads, such as seed heads or a whole section of a design, can really enhance the overall effect. For example, beads were used to replace all of the white cross stitch on the humpback whale and calf design.

To sew on beads, it is best to use a beading needle. Use matching thread and start with the needle at the right side of the fabric. Thread the bead over the needle and on to the thread, then attach it to the fabric by working a half cross stitch. All stitches should run in the same direction so that the beads lie in neat rows on the fabric.

Using Canvas

Many of the projects can be stitched on canvas with tapestry wool (yarn) if you prefer. This would make them perfect for more hard-wearing items such as doorstops, rugs and wall hangings. Ask at your needlework shop for a conversion list to change stranded cotton (floss) colours to wool (yarn). Remember when working on canvas that you will also need a complementary background wool colour to fill in the area around the design.

There is quite a large range of canvas, from lighter weight canvas used for embroidery to heavier canvas used for rugs. Canvas is basically of two types – tapestry and embroidery, and mono and interlock. Tapestry and embroidery canvases are ideal if the design has whole and three-quarter cross stitches or half and quarter cross stitches. Mono and interlock are ideal if the design has all whole cross stitch or half cross stitch.

When altering the count a design is stitched on, remember the design size will change so you need to work out carefully what size canvas is required. The count of any fabric or canvas will tell you how many stitches there are to every 2.5cm (1in). All of the projects have design sizes and stitch counts listed. Simply divide the stitch count by the fabric or canvas count and you will have the approximate size of the design area, without any allowance. Always be generous with the allowances, as you can trim the excess off. When working on canvas, add at least 13–15cm (5–6in) allowance all the way around a design.

Using Embroidery Bands

Embroidery bands, both Aida and evenweave types, are available in many colours, widths and stitch counts and are perfect for decorating many items, such as towels and linens. Aida bands have been used to decorate towels with the turtle and mackerel designs, as follows.

You will need a towel of a complementary colour, plus matching thread for stitching the band on to the towel. I chose a sand-coloured towel for the turtles and a blue-green for the fish. Measure the towel width and add 6mm (¼in) for turnings. Chose your band – I used a 2½in wide, 14 count white Aida band (with a 28 stitch width).

Use the portion of the turtle or mackerel chart on the left and up to the blue arrows (the cut-off line), which will give you a border width of 28 stitches. For a wider band move the cut-off line to the desired number of stitches. Begin stitching 6mm (¼in) from one end of the Aida band (which leaves enough for the turnings). Start at the top of the chart and work to the bottom, repeating the design as many times as necessary for the width of the towel.

When the embroidery is complete, turn the 6mm (¼in) turning ends under, pin, tack, (baste) and machine stitch onto the towel.

Using Waste Canvas

Waste canvas will allow you to work a design on a textured fabric such as fleece or towelling. It is available in various counts and is used just like Aida. It has blue lines running through it to mark off every five blocks making it easier to count stitches.

The waste canvas is tacked over your chosen fabric and is used as a temporary stitching surface, which is then removed to leave the completed design on the fabric.

To use waste canvas, cut a piece at least 5cm (2in) larger than your design. Lay the canvas over the base fabric so the blue lines run vertically along the fabric grain. Pin and tack (baste) in place (Fig 1) and mark the centre.

Fig 1

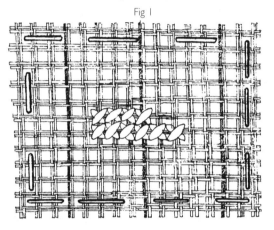

Stitch the design over the waste canvas following the project instructions for the cross stitch. When working, try to ensure that the corners of each stitch share the hole with the previous stitch in the base fabric as this will give a neater finished effect.

When the stitching is complete, remove the tacking (basting) threads, and trim away the excess waste canvas close to the cross stitches. Using a pair of tweezers carefully pull out the vertical threads of the waste canvas – the remaining horizontal threads can then be easily removed (Fig 2). If the waste canvas threads prove stubborn to remove it may help to slightly dampen them.

When all the waste canvas threads have been completely removed, press the embroidery from the wrong side, then add any backstitches, French knots or beads to complete your design.

Fig 2

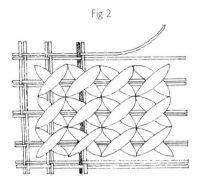

Washing and Pressing Embroidery

If your work has become grubby during stitching, gently hand wash in warm water using a mild liquid detergent. Use a soft nail brush to remove any stubborn marks, rinse in clean water, place the damp fabric on a clean white towel and leave to dry on a flat surface. Do not iron directly on your work as this will flatten the stitches and spoil the finished effect. Lay the work face down on a clean, white towel, cover with a clean, fine cloth and press.

THE STITCHES

Cross stitch embroidery is a simple and straightforward technique and the following section shows you how to work all the stitches used in the book. When following the stitch instructions, please note that stitching is over one block of Aida fabric or two threads of evenweave fabric.

Starting and Finishing Thread

To start off your first length of thread, make a knot at one end and push the needle through to the back of the fabric, about 3cm (1¼in) from your starting point, leaving the knot on the right side. Stitch towards the knot, securing the thread at the back of the fabric as you go (Fig 3). When the thread is secure, cut off the knot.

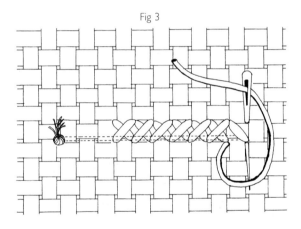

Fig 3

To finish off a thread or start new threads, simply weave the thread into the back of several worked stitches (Fig 4).

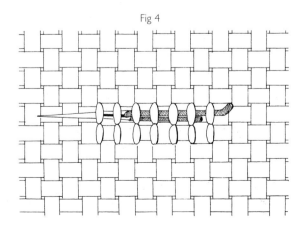

Fig 4

Backstitch

Backstitch is indicated on the charts by a solid coloured line. It is worked around areas of completed cross stitches to add definition or on top of stitches to add detail.

To work backstitch, pull the needle through the hole in the fabric at 1 (Fig 5), and then push back through at 2. For the next stitch, pull the needle through at 3, push to the back at 1, and then repeat the process to make the next stitch. This will give you short stitches at the front of your work and longer stitches at the back. If working backstitch on an evenweave fabric such as Zweigart Brittney, work each stitch over two threads.

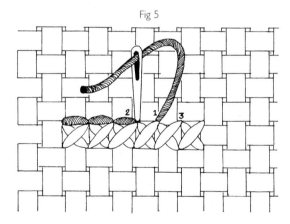

Fig 5

Cross Stitch

Each coloured square on the chart represents one complete cross stitch. Cross stitch is worked in two easy stages. Start by working one diagonal stitch over one block of Aida or two threads of evenweave, then work a second diagonal stitch

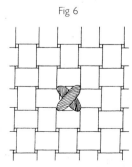

Fig 6

over the first stitch, but in the opposite direction to form a cross (Fig 6).

Cross stitches can also be worked in rows if you have a large area to cover. Work a row of half cross stitches in one direction and then work back in the opposite direction with the diagonal stitches to complete each cross. The upper stitches of all the crosses should lie in the same direction to produce a neat effect (Fig 7).

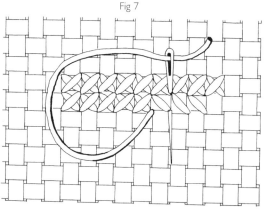

Fig 7

Quarter Cross Stitch

If you chose to work a design on a double mesh canvas using wool (yarn), a quarter cross stitch should be used instead of a three-quarter cross stitch. So, where a three-quarter cross stitch is shown on the chart, use a quarter stitch instead.

To work a quarter stitch, start at one corner of the canvas mesh and work in the same direction as any half stitches, but insert the needle at the corner of the square (Fig 8).

Half Cross Stitch

This stitch is also used if you chose to work a design on canvas in tapestry wool (yarn), replacing whole cross stitches with half stitches. A half cross stitch is simply one half of a cross stitch, with the diagonal facing the

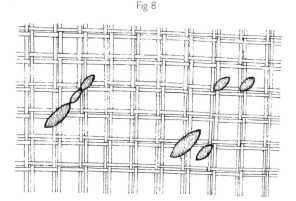

Fig 8

same way as the upper stitches of each complete cross stitch (Fig 8).

Three-quarter Cross Stitch

A small, coloured square taking up a quarter of a chart square represents a three-quarter cross stitch. Forming fractional stitches such as three-quarter cross stitches is less accurate on Aida than on an evenweave or linen fabric because the centre of the Aida block needs to be pierced (Fig 9).

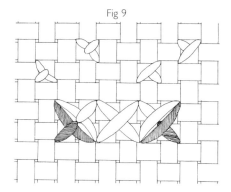

Fig 9

Work the first half of a cross stitch in the normal way, then work the second diagonal stitch in the opposite corner but insert the needle at the centre of the cross, forming three-quarters of the complete stitch. A square showing two smaller coloured squares in opposite corners indicates that two of these three-quarter stitches will have to be made back to back.

French Knots

These are small knots, indicated on the chart by a small coloured dot, which are used to add detail, for example, as eye highlights. Some designs, such as the tawny owl use large areas of French knots and more than one colour.

To work a French knot, bring the needle through to the front of the fabric, just above the point where you want the stitch placed. Wind the thread once around the needle and holding the twisted thread firmly, insert the needle a little away from its starting position (Fig 10).

Fig 10

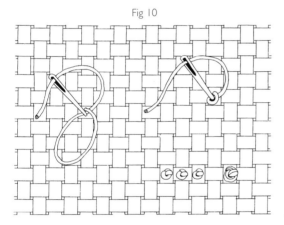

Two tips for working French knots: never rush them and never go back into the same point where your thread came up or your knot will just pull through to the back.

Long Stitch

Long stitches are used to work whiskers and eyebrow hairs and some plant stems and are indicated on charts by a straight, solid coloured line — often in a different, easier-to-see colour. Refer to the instructions for the actual colour. Work long stitches on top of the completed stitched design.

To work long stitch, pull the needle through the fabric at the point indicated on the chart, then push through at the other end, to make a long stitch on top of the fabric. Repeat for the next stitch, carrying the thread across the back of the fabric to the next starting point (Fig 11).

Fig 11

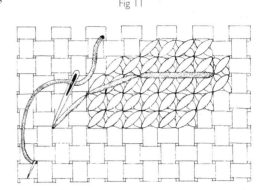

STITCHING TIPS

◆ Steam press your embroidery fabric before stitching to remove any stubborn creases.

◆ Mount fabric onto an embroidery frame or hoop to keep stitches smooth and flat.

◆ Work cross stitches with the top diagonals all facing in the same direction.

◆ Thread up lengths of several colours of stranded cotton (floss) into needles, and arrange these at the side of your work by shade code number or by key reference.

◆ Work the designs from the centre outwards, or split them into workable sections such as quarters. On larger designs, first work the main subject and then complete the background and surrounding designs.

◆ When taking threads across the back of a design, weave the thread through the back of existing stitches to avoid any ugly lines showing through on the right side.

◆ Use short lengths of thread, about 30cm (12in), to reduce any knotting and tangling.

◆ Check your work constantly against the chart to avoid making mistakes.

◆ For a smooth piece of work without any lumps or bumps, avoid using knots at the back, and cut off any excess threads as short as possible.

◆ Keep your work clean by packing it away in its own clean plastic bag to prevent any accidents.

MAKING UP

One of the reasons why cross stitch is so popular, apart from being so easy to work, is that the designs are simple to adapt and use in different ways. Merely by changing the material used — the fabric, the thickness or type of thread or the background colour — a design can be transformed, and examples of this are given throughout the book. The following section describes how to make up the embroideries as illustrated in the book. Most of the designs have been mounted and framed as pictures but there are dozens of other ideas for making up given throughout the Display It pages. When making up any item, a 1.5cm (⅝in) seam allowance has been used unless otherwise stated.

Mounting and Framing

It really is easiest and best to take larger pictures to a professional framer, who will be able to stretch the fabric correctly and cut any surrounding mounts accurately. If, however, you choose to do the mounting and framing yourself you will need a box for cutting mitred edges on frames, some panel pins, a suitable saw, some hardboard (or thick card) and mount board. When choosing mount board and framing, it is best to take your finished work with you, to get the best idea of what the end result will be.

Mount your embroidery onto some thin hardboard or card and fasten in place by lacing it around the card or by stapling it. Decide on the frame size you require and carefully cut your frame pieces to the correct size, then panel pin them together. Using a mount cutter (which is by far the easiest) or a craft or Stanley knife, cut your mount board to the required depth. Place the mount board into the frame, then the embroidery. Finally, cut hardboard to size for the backing and wedge in with metal clips or tape in place.

Using Ready-made Items

Many of the projects in the book can be displayed in ready-made items such as trinket boxes, stools, mugs, teapot stands, coasters, cards and small frames or flexi-hoops (see Suppliers). Smaller pieces of embroidery can be backed with lightweight iron-on interfacing to prevent the fabric wrinkling, and then mounted following the manufacturer's instructions.

Making a Cushion with a Frill

Many of the designs in the book can be made up into gorgeous cushions. The instructions which follow are for a 38 x 38cm (15 x 15in) cushion with a frilled edge but the principles can be applied to any size. You will need a cushion pad, matching sewing thread and sufficient fabric for backing and making a frill.

For the cushion front, carefully cut away the excess embroidery fabric, leaving a square with a 1.5cm (⅝in) seam allowance. For the cushion backs, cut two pieces of cotton fabric 25.5 x 38cm (10 x 15in). Take each rectangle and hem along one long edge. To make the frill, cut enough 12cm (4¾in) wide strips along the length of the cotton fabric to give a finished length of 2.9m (3⅛yd). With right sides facing, stitch the fabric strips together along the short edges to form a circle. Press seams open. Fold the strip in half so that the long edges meet, enclosing the raw edges of the short seams, then press (Fig 12).

Run two rows of gathering threads along the raw edges of the frill fabric, then pull up the thread until the frill is the right length, distributing the gathers evenly. With the embroidered fabric facing outwards, place the frill around the outer edge of the cushion front, so raw edges face outwards. Distribute the gathers evenly, then pin, tack (baste) and machine stitch in place.

Lay the embroidered fabric face upwards on a flat surface. With right sides down, lay the two rectangles of the cushion back on top of the front so that all the raw edges match and the hemmed edges overlap at the centre. Pin, tack (baste) and then machine stitch along the stitching line and through all layers of fabric. Neaten the raw edges, then turn the cover through to the right side and insert a cushion pad.

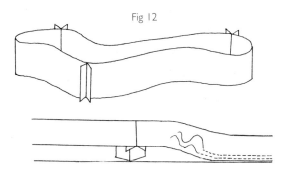

Fig 12

Making a Cushion with a Braid Edge and Tassels

This type of cushion makes an elegant statement and could be used to display many of the designs, including the large collages such as the Safari Collection and also smaller designs like the little bee-eaters. A

pincushion can also be made following the same principles. You will need a cushion pad, matching sewing thread, sufficient fabric for backing and enough thick furnishing braid to go around your cushion. Follow the instructions for making the frilled cushion, above, but omit the frill.

After making the cushion cover, turn it through to the right side. Measure the length of the cushion edges, then cut lengths of braid to the same measurements, adding 10cm (4in) to each length. Hand stitch the braid along each edge, leaving equal lengths extending at each end. Pinch together the two lengths of braid extending from each corner, then use matching sewing thread to tightly bind them together. Secure the sewing thread with a knot, then fray the braid to make a tassel (Fig 13).

Fig 13

Making a Doorstop

Making work up as a doorstop is easy to do and would make a wonderful feature of such designs as the tawny owl and the puma. You will need medium-weight interfacing, cotton backing fabric, a quantity of wadding (batting) and some clean kitty litter or sawdust.

Back the embroidery with a medium-weight iron-on interfacing, following the manufacturer's instructions. Use a soft pencil to draw a line all around the design 1.5cm (⅝in) from the finished embroidery and cut away excess fabric along this line. Cut a piece of cotton backing fabric to the same shape.

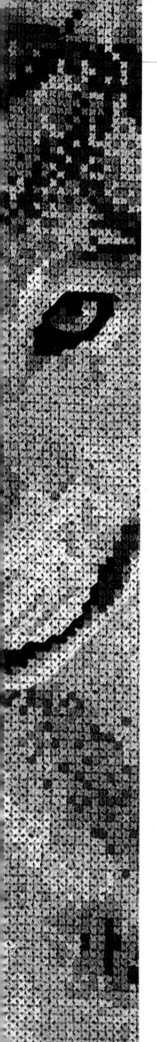

With right sides facing, place the front and back pieces together then pin and tack (baste) around the edges. Machine stitch the layers together, taking a 1.5cm (⅝in) seam allowance and leaving a 20cm (8in) gap along the bottom straight edge for turning.

Turn the doorstop through to the right side and almost fill with wadding (batting). Then, fill a sock with kitty litter or sawdust and tie a knot at the end. Insert the filled sock into the base of the doorstop then close the gap with slipstitches.

Making a Wall Hanging or Rug

Wall hangings and small rugs are lovely ways to display embroidery. Wall hangings can be stitched in either stranded cotton (floss) or wool (yarn). Rugs need to be more hard wearing and so need to be stitched on canvas with wool (yarn). See page 119 for tips on working with canvas.

To convert the wolf design into a wool wall hanging, I would suggest using a 10 count embroidery/tapestry canvas, as the design has three-quarter cross stitches. You can stitch it with one thread of tapestry wool (yarn) either in whole and three-quarter cross stitch or in half and quarter cross stitch.

The puma design would look wonderful stitched up as a rug, as it is worked using whole cross stitch. Use a 7 count mono or interlock canvas, whole cross stitch and two threads of tapestry wool (yarn).

EDGING RUGS AND WALL HANGINGS
The edges of the finished embroidered design will need to be neatened and strengthened. Trim away excess canvas to within 10cm (4in) of the stitches. Fold the turnings to the back, leaving one hole showing and one thread running across the top of the fold, then stitch the turnings in place by stitching an overcast edging using two strands of the desired colour. Secure the thread at the back of the canvas, insert the needle into the hole nearest the embroidery stitches and pull to the front. The overcast thread should share a hole with the last stitch of the design. Take the needle to the back of the canvas and work the next stitch in the same way. Work one, two or three stitches in each hole to cover the canvas completely (Fig 14).

Fig 14

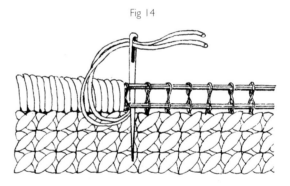

HANGING A WALL HANGING Either sew on curtain rings at intervals along the top on the wrong side, at a depth of 1½–2in (4–5cm), or make loops by cutting two 20cm (8in) strips of flat braid. Fold the strips in half to form a loop, then use sewing thread to hand stitch them in place at the back of the wall hanging at either end of the top edge. Slide a curtain rod or length of wooden dowelling through these to suspend it from the wall, from cup hooks or other suitable fastenings.

SUPPLIERS

If you should require any further information about products, catalogues, price lists or local stockists from any of the suppliers mentioned, contact them direct by post or phone. Remember to always include a stamped addressed envelope. If contacting them by phone, they will be able to tell you if there is any charge for the catalogue or price lists.

DMC Creative World

Pullman Road, Wigston, Leicester LE18 2DY. Tel: (0116) 281 1040.
For all threads, embroidery fabrics and beads used throughout the book, and for the name and address of your nearest DMC and Zweigart stockist.

DMC threads are supplied in the USA by:
The DMC Corporation, South Hackensack Ave, Port Kearny, Building 10A, South Kearny, NJ 07032-4688. www.dmc-usa.com
Zweigart fabric is supplied in the USA by Joan Toggitt Ltd,
2 River View Drive, Somerset, NJ 08873-1139.
E-mail: info@zweigart.com
www.zweigart.com

Framecraft Miniatures Ltd

376 Summer Lane, Hockley, Birmingham B19 3QA.
Tel: (0121) 2120551.
For ready-made items for embroidery. Suppliers of the hexagonal teapot stand, coffeepot stand, mugs and cards.

Framecraft products are also supplied worldwide by:
Anne Brinkley Designs Inc,

761 Palmer Avenue, Holmdel, NJ 97733, USA.
Gay Bowles Sales Inc, PO Box 1060, Janesville, WI 53547, USA.
Ireland Needlecraft Pty Ltd, 4, 2–4 Keppel Drive, Hallam, Vic 3803, Australia.

Market Square (Warminster) Ltd

Wing Farm, Longbridge Deverill, Warminster, Wilts BA12 7DD.
Tel: 01985 841042.
Suppliers of the chest for the turtle design.

Sleepy Hollow

Duke House, Duke Street, Kingsbridge, Devon TQ7 1HU.
Tel: 01548 857949.
Suppliers of the smock tops for the zebra and elephant designs.

Vilene products were used on projects throughout the book. A selection of iron-on interfacings is available in major department stores and all good haberdashery shops (notions departments).

ACKNOWLEDGMENTS

A very special thank you to my husband Ian, for all his support, help and understanding while I worked on this book. Thank you also goes to Pops, who always reminds me how much I love my work when I'm feeling under pressure and over-worked, and to my mother and my two brothers, Nick and Edward, who are always there with advice and encouragement when I need it.

Thanks to Doreen Montgomery for her encouragement, advice and support from the outset of this book. Thank you also to the following people at David & Charles for their contributions and help with getting this book published: Cheryl Brown for her persuasion, encouragement and advice from the beginning; my editor Linda Clements for her invaluable help with the text; and also Ali Myer, Lisa Forrester and Sandra Pruski. Thanks to Jon Bouchier for the wonderful photography, and to David Lynch for the photograph on the back cover flap.

Thank you also to Cara Ackerman at DMC and Sarah Carlton-Gray at Framecraft Miniatures Ltd for their support. A huge thank you to John Parkes of Outpost Trading, who, as usual, had to rush my framing through at the last moment. Thank you to Sleepy Hollow for letting me use their wonderful smock tops for the zebra and elephant designs.

Index

Charts are in *italic*   Chapters headings in **bold**

Index